From Marginal Adjustments *to* Meaningful Change

Rethinking Weapon System Acquisition

John Birkler, Mark V. Arena, Irv Blickstein,
Jeffrey A. Drezner, Susan M. Gates, Meilinda Huang,
Robert Murphy, Charles Nemfakos, Susan K. Woodward

Prepared for the Office of the Secretary of Defense
Approved for public release; distribution unlimited

 NATIONAL DEFENSE RESEARCH INSTITUTE

The research described in this report was prepared for the Office of the Secretary of Defense (OSD). The research was conducted within the RAND National Defense Research Institute, a federally funded research and development center sponsored by the Office of the Secretary of Defense, the Joint Staff, the Unified Combatant Commands, the Navy, the Marine Corps, the defense agencies, and the defense Intelligence Community under Contract W74V8H-06-C-0002.

Library of Congress Cataloging-in-Publication Data

From marginal adjustments to meaningful change : rethinking weapon system acquisition / John Birkler ... [et al.].
    p. cm.
  Includes bibliographical references.
  ISBN 978-0-8330-5046-5 (pbk. : alk. paper)
   1. United States. Dept. of Defense—Procurement. 2. United States—Armed Forces—Procurement. 3. United States—Armed Forces—Weapons systems.
  4. Defense contracts—United States. I. Birkler, J. L., 1944-

  UC263.F758 2010
  355.6'212—dc22

                                                            2010042130

The RAND Corporation is a nonprofit institution that helps improve policy and decisionmaking through research and analysis. RAND's publications do not necessarily reflect the opinions of its research clients and sponsors.

RAND® is a registered trademark.

Cover design by Eileen Delson La Russo

Published 2010 by the RAND Corporation
1776 Main Street, P.O. Box 2138, Santa Monica, CA 90407-2138
1200 South Hayes Street, Arlington, VA 22202-5050
4570 Fifth Avenue, Suite 600, Pittsburgh, PA 15213-2665
RAND URL: http://www.rand.org/
To order RAND documents or to obtain additional information, contact
Distribution Services: Telephone: (310) 451-7002;
Fax: (310) 451-6915; Email: order@rand.org

Preface

In today's defense environment, pressure is growing on policymakers to make the defense acquisition system more nimble and effective. To help, prior to the 2008 U.S. presidential election, the RAND Corporation was asked to prepare a series of white papers as part of the Office of the Secretary of Defense's effort to provide the next administration with guidance on defense acquisition challenges in several areas likely to be of critical importance to the new defense acquisition leadership: competition, risk management, novel systems, prototyping, organizational and management issues, and the acquisition workforce. These efforts led to six occasional papers that offer thought-provoking suggestions based on decades of RAND and other research, new quantitative assessments, a RAND-developed cost-analysis methodology, and the expertise of core RAND research staff. The papers generated considerable interest; indeed, demand exhausted the initial print run. This monograph, a compilation of those six papers, will be of interest to members of the acquisition and military requirements communities.

This research was sponsored by the Office of the Under Secretary of Defense for Acquisition, Technology, and Logistics (OUSD (AT&L)) and conducted within the Acquisition and Technology Policy Center of the RAND National Defense Research Institute, a federally funded research and development center sponsored by the Office of the Secretary of Defense, the Joint Staff, the Unified Combatant Commands, the Navy, the Marine Corps, the defense agencies, and the defense Intelligence Community.

For more information on the RAND Acquisition and Technology Policy Center, see http://www.rand.org/nsrd/about/atp.html or contact the director (contact information is provided on the web page).

Contents

Figures

Tables

Summary

Despite years of change and reform, the Department of Defense (DoD) continues to develop and acquire weapon systems that it cannot afford and cannot deliver on schedule. Consequently, defense acquisition is one of the most urgent issues that DoD has to address today—a point emphatically conveyed by Deputy Secretary of Defense William Lynn during his confirmation hearing: "[A]cquisition reform is not an option, it is an imperative" (2009, p. 10).

This monograph is designed to inform new initiatives for markedly improving the cost, timeliness, and innovativeness of weapon systems that DoD intends to acquire. It is the result of a RAND effort that led to six occasional papers on topics that are likely to be of critical importance to DoD leadership: competition, novel systems, prototyping, risk management, organizational and management issues, and the acquisition workforce. These papers build on RAND staff's deep experience in acquisition management issues to provide innovative ideas and suggestions to revitalize defense acquisitions.

Findings

Savings from Competition Are Not Inevitable

The value of competition is so much taken for granted that defense officials are often criticized for not relying more frequently on competition in awarding contracts for major defense systems. However, a second production source does not guarantee savings in every procurement. Defense acquisitions differ from the typical business market in terms of priorities, the number of buyers and producers, and the level of market uncertainty. Moreover, competition requires additional time, money, and management effort.

RAND researchers used historical data and a RAND-developed methodology to determine whether and when competition is a reasonable acquisition strategy during the production phase. The analysis indicates that competition is more reasonable in situations in which nonrecurring costs are low, cost improvement is minimal, and a greater number of units will be produced. In some cases—especially in the procurement of major systems whose nonrecurring costs are large—it may actually be less costly for the government to forgo competition.

DoD Must Accept More Risk to Meet Demand for Novel Systems

Today, there is a growing need to respond to asymmetrical threats using novel weapon systems that can be quickly developed and fielded. Novel systems—such as the F-117 Stealth Fighter and robotic ground vehicles—involve more uncertainty than conventional systems, not only with regard to design and technology but also in terms of how they will be used, how many units will be needed, and how much they will cost (see Table S.1).

Current acquisition policies and processes are too risk averse to enable the effective development and timely employment of novel systems. Consequently, DoD needs a separate acquisition strategy that

Table S.1
Comparison of Conventional and Novel Systems

Dimensions	Conventional Systems	Novel Systems
Design	Follow-on	New
Technology	Evolutionary	Disruptive
Operational employment	Established	In formulation
Outcomes	Predictable	Uncertain
Production run	Large	Uncertain
Operational life	Long	Uncertain

is less tied to achieving precise cost, schedule, and performance outcomes. The new strategy should include a focus on unique integrations of existing and emerging technologies, a willingness to accept risks, easy and quick termination of programs not yielding expected benefits, and early test and demonstration of military utility.

Oversight Is Based on Dollar Value, Irrespective of Risk

DoD assigns responsibility for decisions on major defense acquisition programs on the basis of the program's dollar value. The higher the value, the more senior the decisionmaker. This approach has been constantly refined over the years without having noticeably improved acquisition outcomes. A new paradigm in which the level of oversight and management would be based on the level of risk a program represents would help DoD more effectively manage weapon system programs. Some very costly projects might have significantly less risk than projects of similar cost and thus should require less oversight. Conversely, projects may cost little but have a lot of risk because they push the state of the art in technology; such programs require more-comprehensive oversight than dollar value alone would indicate.

Cost, schedule, and performance are the primary attributes by which programs are assessed, but more-discrete program attributes—such as technical, system, design, production, and business innovation risk—would better enable program managers to look ahead and act to avoid adverse outcomes. The Defense Acquisition Management System has sufficient tools and allows time for conducting proper assessment and management of technical risk and, to some extent, system integration risk. However, new approaches in design, production, and business areas of acquisition programs do not appear to receive the same level of skepticism and comprehensive oversight received by new technologies and systems. Descriptive levels of risk that could be used to assess new design approaches include the following:

- New, unproven processes. New design tools under development. New design organization.
- Large expansion of existing design organization. Many new designers and supervisors unfamiliar with design tools and processes.

- Existing design organization using radically changed design tools, processes, and/or technologies.
- Experienced design organization using new design tools with proven processes.
- Experienced design organization using existing, proven design tools and processes.

Organizational Schisms and Rigid Processes Contribute to Inefficiencies

Many of the problems that contribute to poor cost and schedule outcomes are systemic to the way that the acquisition process is organized and managed in DoD. Specifically, organizational schisms and overly prescribed management processes contribute both to inefficiencies in the acquisition system and to unrealistic expectations.

For example, service chiefs[1]—who validate warfighting requirements—have become increasingly disconnected from the service acquisition executives who develop and acquire new weapon and information systems in conjunction with their program executive officers and program managers. Without sufficient dialogue between these entities, service chiefs may emphasize warfighting needs at the expense of reducing cost, and the acquisition process loses their operational insight, which is critical in analyzing trade-offs between cost, schedule, and performance. The service chiefs should have a central, but not controlling, voice in the acquisition process to enable the requirements, funding, and acquisition processes to function together well. Having the service vice chiefs serve as co-chairs of the military departments' acquisition boards would be a step in the right direction. However, increasing the role of the combatant commands in these decisionmaking processes would require them to spend too much time away from their warfighting responsibilities.[2] It is the job of those in the Pentagon to reach out to the combatant commands and demonstrate that their needs are being addressed.

---

[1] That is, the Chief of Staff of the U.S. Air Force, the Chief of Staff of the U.S. Army, the Chief of Naval Operations, and the Commandant of the Marine Corps.

[2] The military departments supply forces to the combatant commands, which conduct joint military operations.

Recent acquisition reforms have made management processes overly complex and rigid, leading to an environment in which "success" is measured by an ability to follow rules in a rote manner to move a program through an increasing number of wickets. DoD needs a more streamlined requirements and acquisition process, one that, unlike the current process that prescribes everything through an instruction or regulation, encourages workforce initiative and responsibility.

Evidence of the Benefits of Prototyping Is Mixed

Acquisition policy and practice reflect the recurring theme that prototyping as part of weapon system development can reduce cost and time; allow demonstration of novel system concepts; provide a basis for competition; validate cost estimates, design, and manufacturing processes; and reduce or mitigate technical risk. A review of four decades of RAND research on prototyping indicates that the available evidence on its benefits is somewhat mixed overall. Nevertheless, the historical record does reveal some of the conditions under which prototyping strategies seem most likely to yield benefits in a development program. These include ensuring that prototyping strategies and documentation are austere, not committing to production during the prototyping phase, making few significant design changes when moving to the final configuration, and maintaining strict funding limits.

Existing case studies and statistical analyses present the policymaker with mixed results, so, in essence, DoD's new competitive prototyping mandate was incorporated into policy without a strong link between the new policy emphasis and its intended improvements to program cost, schedule, and performance outcomes. A carefully structured analysis of prototyping strategies emphasizing recent experiences with competitive prototyping (with, e.g., F-22 fighter aircraft, the Joint Strike Fighter, the Littoral Combat Ship) would help ensure a more successful implementation of the new policy.

DoD Lacks Systematic Data on the Acquisition Workforce

Through the end of FY 2015, DoD plans to increase the defense acquisition workforce by 20,000 workers (16 percent), converting contractor positions to civil service positions and hiring new civil servants. This step responds to three common claims: (1) The acquisition workforce is too small to meet current workload, (2) it lacks the necessary skills, and (3) contractors are overused or inappropriately used to perform acquisition functions. However, DoD does not have systematic data on workforce supply and demand, the adequacy of workforce skills, or the amount and nature of contractor support. Without such data, it is difficult to determine whether and to what extent workforce attributes affect acquisition outcomes.

To gain insight into DoD's acquisition workforce in terms of supply and demand, a RAND analysis drew upon data about the department's overall civilian workforce. These data indicate that the number of DoD civilians in acquisition-related occupations declined during the 1990s, reaching a low of 77,504 in 1999, and then climbed steadily to reach 119,251 in 2005. By 2006, it had been reduced slightly to 113,605 (see Figure S.1). The greatest declines occurred in contracting, quality assurance, and auditing—groups that were the most likely to have been affected by increased workload due to procurement reforms and increased use of contractors.

Figure S.1
Civilians in the Acquisition Workforce, September 30 Annual Snapshots

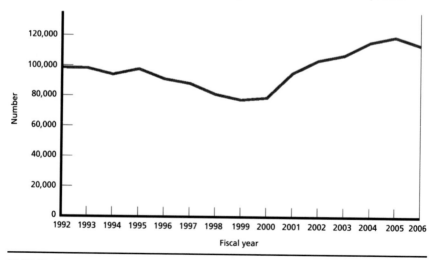

SOURCE: Gates et al., 2008, Figure 3.1.

DoD should acquire the evidence needed to make the case for workforce changes by gathering data on the total workforce, including contractors, mapping the workforce to acquisition activities for which performance may be measurable, and identifying and gathering information on processes and outcomes that the workforce can be expected to influence.

Conclusions

The following chapters contain more-detailed proposals to improve defense acquisition through initiatives focused on competition, novel systems, risk management, organizational factors, prototyping, and the acquisition workforce. The starting point for these proposals is the following list of overarching conclusions:

- Explicit evaluation of the pros and cons of production competition should be undertaken for each acquisition.
- The characteristics of novel systems are so different from those of the systems for which the present acquisition process was designed that they require a separate acquisition strategy.
- Managing defense acquisition programs by risk, rather than dollar amount, merits serious consideration.
- Some bold steps are needed to clear structural impediments to acquisition reform. Greater participation of the service chiefs in a more streamlined acquisition system would most closely align requirements with contracting for material and material support.
- The successful application of prototyping strategies in the future requires certain conditions, such as testing critical performance attributes in a realistic environment.
- DoD needs to invest more in understanding the strengths and weaknesses of the acquisition workforce, including contractors.

Abbreviations

AAC	Air Armament Center
ACAT	acquisition category
ACTD	Advanced Concept Technology Demonstration
AFMC	Air Force Materiel Command
AMRAAM	Advanced Medium-Range Air-to-Air Missile
AOR	Area of Responsibility
ASDS	Advanced SEAL Delivery System
ATD	Advanced Technology Demonstration
AW	acquisition workforce
CBO	Congressional Budget Office
CDR	critical design review
CDS	Combat Data System
CINC	commander in chief
COCOM	combatant command
DAWIA	Defense Acquisition Workforce Improvement Act
DoD	Department of Defense
DoD IG	Department of Defense Inspector General
DPAP	Defense Procurement and Acquisition Policy

EDM	engineering development model
EMD	engineering and manufacturing development
EMRL	Engineering and Manufacturing Readiness Level
FFRDC	federally funded research and development center
FRP	full-rate production
GAO	Government Accountability Office
HCSP	Human Capital Strategic Plan
IOT&E	initial operational test and evaluation
IPL	Integrated Priority List
IPPD	Integrated Product and Process Development
IRL	Integration Readiness Level
JCIDS	Joint Capabilities Integration and Development System
JCTD	Joint Concept Technology Demonstration
JSF	Joint Strike Fighter
LCS	Littoral Combat Ship
LRIP	low-rate initial production
LWF	Lightweight Fighter
MDA	Milestone Decision Authority
MDAP	Major Defense Acquisition Program
NASA	National Aeronautics and Space Administration
NAVSEA	Naval Sea Systems Command
NSPS	National Security Personnel System
NVR	Naval Vessel Rules

OMB	Office of Management and Budget
OPM	Office of Program Management
OSD	Office of the Secretary of Defense
OUSD (AT&L)	Office of Under Secretary of Defense for Acquisition, Technology, and Logistics
PDR	preliminary design review
PPBE	planning, programming, budgeting, and execution
R&D	research and development
RAN	Royal Australian Navy
RCR	required cost reduction
RFP	request for proposal
S&T	science and technology
SAR	Selected Acquisition Report
SPO	system program office
SPRDE	systems planning, research, development, and engineering
SRL	System Readiness Level
TDP	technical data package
TDS	Technology Development Strategy
TRL	Technology Readiness Level
UAS	unmanned aerial system
UH	utility helicopter
UID PMO	Unique Identification Program Management Office
USN	United States Navy

CHAPTER ONE

Determining When Competition Is a Reasonable Strategy for the Production Phase of Defense Acquisition

Mark V. Arena and John Birkler

Introduction

The use of competition in weapon system acquisition is widely advocated in policy statements and widely reflected in requirements issued by Congress, the Office of Management and Budget (OMB), the Department of Defense (DoD), and the military services. This emphasis stems from the conviction that competition during the production phase of the acquisition system will drive the unit cost of a system or subsystem down and reduce overall procurement cost to the government.[1] Other arguments for having more than one producer exist (e.g., providing a surge capability should the services need to expand production quickly), but the crux of the competition issue is procurement cost (or, more accurately, price).

> It is not self-evident that a second production source will produce savings—especially when nonrecurring costs are large.

We do not question the value of competition as a means of inducing a firm to reduce prices, but it is not self-evident that a second production source will produce savings for the government in every procurement. In some cases—especially in the procurement of major systems where the nonrecurring costs are large—it may actually be less

---

[1] A variety of competitive strategies exist. Each strategy addresses particular features of major acquisitions and is applicable to different phases of the acquisition process. The focus of this analysis is on saving money during the production phase.

costly for the government to forgo competition and to rely on a single supplier.

Senior DoD officials must determine whether competition is likely to result in savings or losses for the government. This paper compares the characteristics of a typical business market to those of defense acquisitions, identifies the benefits and drawbacks of competition in defense acquisitions specifically, and shows how to determine when the introduction of competition during production is a *reasonable* acquisition strategy.

Pentagon Acquisitions: Not Business As Usual

The complexity and uniqueness of major defense procurement make it difficult for DoD to follow typical commercial business price-competition approaches. In the typical business market, a buyer examines the available products, requests competitive bids for production from a number of contractors, selects a bid based on a fixed price, and signs a one-step contract for delivery on a specified date. Such a market depends on having complete information about a customer's needs; a standardized, off-the-shelf product; a predictable budget; certainty about the number of items to be purchased; and little reason for concern about the future viability of the losing firm. Major defense acquisitions lack these characteristics.

If the unique characteristics of defense procurement are not taken into account, expectations for the use of price competition may be unrealistic.

In major defense acquisitions, the relationship between buyer and producer is almost completely different from that assumed in the economist's model of the typical business market. For example, defense acquisitions have only one domestic buyer; producers typically compete during the design stage as opposed to the production stage; and concern about sustaining a unique industrial sector factors into the buyer's decisionmaking process. Table 1 summarizes the characteristics assumed for the typical business market (or "perfect market") and compares them to the corresponding characteristics of typical major defense acquisitions during production.

As indicated in the table, quality and schedule are often more important criteria than price in defense acquisitions. The high priority the defense buyer usually gives to product quality is sometimes regarded as a defect in the acquisition process. The conventional wisdom is that when programs experience difficulties, expenditure is the first constraint to be relaxed and schedule the second, but that performance goals are adhered to quite rigorously, with the result that the unit price of the product increases.

The data support this description of the way quality, schedule, and price are traded off, but it is by no means clear that this should be accepted as adverse criticism. The Services' emphasis on high system quality is consistent with the long-established national policy that emphasizes quality rather than quantity for defense, and hence calls for the development and production of systems superior to those fielded by possible opponents. If this emphasis is accepted, the question is not so much whether quality should be given priority, but rather, what kind of quality and how much quality is enough? This latter question, however, is peripheral to what concerns us here. It is sufficient to recognize that major system acquisitions generally aim at a quality of product that requires innovations in design and the application of advanced technologies, with all the technical uncertainty this entails.

Comparison of the typical business market and defense acquisitions helps illustrate the inherent difficulties in introducing effective price competition into defense acquisitions. Unless these differences are taken into account, expectations for the use of price competition in defense purchases may be unrealistic.

The Benefits and Drawbacks of Competition

The basic argument for competition in defense procurement is that it is believed to reduce the government's costs of purchasing goods and services. Nonetheless, in some cases, it may actually be less costly for

Table 1.1
Characteristics of "Perfect Market" Versus Typical Characteristics in Major Defense System Production

Perfect Market	Major Defense System Market
Many buyers and producers, none being dominant; each buyer has a choice of many producers.	*Only one buyer. Usually only one producer—the prime contractor who developed the system.*
To a close approximation, price (a firm-fixed price) is determined by the "hidden hand" of the market.	*Production prices (seldom truly firm-fixed prices) are determined by a series of negotiations in a sole-source environment.*
Product is an existing, standardized item, similar for each producer—it is "homogeneous," and its characteristics are stable over time.	*Product is a newly developed item, usually without close substitutes and with a design that continues to change during much of the production phase and often even afterward.*
Competition focuses primarily on price, but other criteria (such as quality, reliability, and performance) are considered.	*Prospective producers compete early in the development phase through "design rivalry." Buyer is concerned with product quality (especially performance), delivery schedule, and other nonprice factors. Price is not the dominant consideration in selecting the producer; quality of the product is normally given priority.*
No producer has an advantage in production technology or economies of scale.	*Production technology is dynamic and may differ among prime contractors and their subcontractors. Economies of scale, including "learning-curve" and production-rate effects, significantly influence producer costs. A superior developer is not necessarily a more efficient producer.*
Market is easy for new producers to enter.	*New prime contractors seldom enter the defense sector—entrance is inhibited by the high capital investment required, the proprietary rights of others, and the administrative and contractual burdens of a highly regulated industry.*

(continued)

Table 1.1—Continued

Perfect Market	Major Defense System Market
Buying the product is a simple, quickly completed, one-step transaction between the buyer and the producer, independent of other purchases from the same or other producers.	*Acquiring a major system is a multiyear, multistep, complex process, involving scores of successive and usually interdependent contract negotiations between buyer and producer.*
Market is characterized by near perfect intelligence and absence of uncertainty. Information about product price, standards of quality, number of items purchased, and delivery schedule is freely available to all concerned.	*Uncertainty is a dominant and largely unavoidable feature. Among the market uncertainties are the* • *threat the system will face* • *most suitable system capabilities* • *best design approach* • *feasibility of development* • *time and other resources required to complete development and make the transition to production* • *deficiencies that may be revealed by operational testing.*

the government to forgo competition during the production phase and rely on an alternative strategy. In this section, we discuss the general benefits of competition as well as common drawbacks for weapon system acquisition, including time; money; management effort; and a long-term, uncertain benefit.

Benefits of Competition

The value of competition in U.S. society is so much taken for granted that defense procurement officials are often criticized for not relying more frequently on head-to-head competition in awarding production contracts for major defense systems. Those critics argue that competition produces many significant benefits. Competition improves product quality and lowers unit costs, they say, compared with a noncompetitive environment. Competition

Competition can save money, improve product quality, ensure equity, and protect the industrial base.

forces manufacturers to quickly learn about new technologies and production techniques, fostering greater technological progress and industrial productivity. Finally, competition allows for a more equitable process under which acquisition contracts are awarded. The government has the responsibility to treat firms fairly; competition allows companies that believe they can make a competitive bid to do so, which makes the bid process more fair than sole-source procurement.

We do not question the value of competition as a means of inducing a firm to reduce prices. When competition or the threat of competition is perceived as real, a firm may act in a number of ways to cut costs and price. Managers will often assign their best people to a competitive program, allocate corporate capital for equipment, and fund value-engineering studies (rather than expecting the customer to fund them). A company may also transfer production from an area of high labor costs, such as California or Massachusetts, to locales where labor costs are lower. Management may take measures to substitute capital for labor, accelerate cost-reduction schemes, and seek out alternative vendors. A firm may be able to operate at an economical rate by producing enough parts in a few months to satisfy the contractual requirement for an entire year, and then assign the workers to other tasks for the remainder of the year. In addition, a company is usually able to reduce the number of engineering and manufacturing support personnel assigned to a program. Noncompetitive programs tend to be heavy in such personnel, often because the customer wants to retain the services they provide.

Saving money is not the only benefit of bringing a second producer into a program. Improved quality assurance is often cited as a reason for introducing a second production source. In some instances, the underlying reason for a second source has been a profound dissatisfaction with the initial contractor, which may be a good developer but an inefficient producer. The nature of defense procurement is such that, once a contractor is chosen to develop a major new system, the responsible military service is locked into a relationship with that contractor that could last 20 years or more. Bringing a second company into a program is an effective way to encourage greater cooperation from the initial firm.

Drawbacks of Competition

Barriers to competition in defense acquisitions also exist. Analysts note that competition requires additional time and money and also entails extra management complexity and effort. At the same time, most of the benefits of competition are long term, providing program managers with little incentive to implement competitive steps because payoff is well in the future. Further, competition has uncertain and mixed results. In a risk-averse environment, this uncertainty also reduces the program manager's incentive to use competition.

We discuss each drawback separately below. Throughout the discussion, we attempt to distinguish among different problems that arise during discrete phases of the acquisition cycle.

Additional Time and Money. At almost every phase in the acquisition cycle and for almost every kind of competition, adding a second competitor requires current-year investment above what a sole-source would cost. During the planning phase, such funds are relatively small in absolute terms. However, when the program moves to the production phase, the magnitude of the funding required for a second, competitive source becomes large relative to program costs and in absolute terms, reaching tens or hundreds of millions of dollars.

Such a funding commitment can be difficult to obtain. It will have to come from higher up the chain of command, which means that many people will have to be "sold" on the competitive action. At every level in the organization, there will be some who are sympathetic to the request for funds and others who will see themselves as competing for the same funds. Some groups will tend to underestimate the difficulty of developing a particular system or have an interest in fielding it very quickly, thus they will resist competition during full-scale development on the grounds that it wastes time and money. The situation is even more complex in multiservice programs, wherein all the armed services must agree to put up the extra money.

> **Funding for a second source is substantial; it can be difficult to secure and maintain throughout a development program.**

When substantial amounts of money are involved, DoD and multiple congressional committees must be sold on the competition as well. When there is no great pressure for competition and when other acquisition initiatives are being emphasized, DoD and the congressional committees involved can be difficult to convince. Congress tends to dislike programs with heavy front-end cost; other, less obvious political problems sometimes intrude as well. Funding requests are reviewed by four different congressional committees, which do not automatically coordinate their decisions, so each must be persuaded separately. It is not unusual for one committee to support a competition and another to delete the funds for it. DoD and Congress can also hold strongly differing positions.[2]

Further, once funding for a competition is approved, there is no guarantee that it will be maintained. Money for competitive programs is a prime target in a budget squeeze, and initial high-level support for competition may evaporate. In the services and in DoD, there are frequent changes in top-level personnel; when new people take over, they inevitably change priorities. Written policy supporting competition remains fairly consistent, but interest in competition changes with personnel. The result is that it can be difficult to maintain all the funding necessary to conduct a competitive program.

An additional barrier to competition is the time involved in testing or qualifying a second contractor. Schedules can also lengthen due to adjudication of protests by the losing firm and because competition can increase program complexity and bureaucratic involvement. By lengthening schedules, competition carries the risk of raising program costs. The risk of increased program length is a disincentive to competition because there is usually a strong desire to deploy the system as rapidly as possible.

---

[2] For example, the services have strongly fought against having a second General Electric engine developed to compete against the Pratt & Whitney engine for the F-35, but Congress continues to insist on funding the General Electric engine in order to maintain competition.

Extra Management Effort. Competition increases the workload of the project office. This extra work stems from several sources: additional planning, qualifying a second producer, and quality control and configuration management issues. If a competition is to be beneficial, considerable planning for the competitive steps is necessary. The request for proposal (RFP) must be prepared and the source selection process must be designed. The program office must comply with regulations designed to ensure the fairness of the competition. This process involves special security to deal with "competition sensitive" material, special reports, et cetera. Competition also introduces the possibility of lawsuits, disputes, and charges of unfairness by the contractors who lose the competition. Consequently, the source selection must be carried out in a way that not only chooses the best design, but also raises a minimum number of questions about fairness. In addition, if awards are granted to more than one contractor, each additional contractor the program office must deal with usually means more work. This is especially true when cost-type contracts are involved because the program office must monitor the costs of each contractor.

> **Competition involves additional planning, extra work to qualify the second producer, and difficulties in configuration management.**

Competition during production can introduce more management complications and can be a major effort, especially when qualifying a second producer after production has begun. It is difficult and expensive to get a good technical data package (TDP) for the second contractor to use in starting production, and even more difficult to persuade the first producer to pass along to a competitor the benefits of its manufacturing experience.[3] Program managers can choose to develop their own TDP, but for major programs this is almost impossible. Not all services have in-house capability to evaluate a TDP, and

---

[3] There are various levels of detail for a TDP that must be validated, at government expense, by a competitive producer before submitting a final bid for production. An alternative to using a TDP is the form-fit-function (FFF) approach for the second-source product.

without this capability, it is difficult to judge the adequacy of a TDP. Even with a good TDP, it frequently takes a major effort by the program office to help the second source through all its technical problems and into production. In some cases, the second source never succeeds in producing a usable product. Even in those cases where the second source is not successful, the pressure on the first contractor may still make the effort worthwhile.

Another source of additional work in developing a second source is that the program office must work with both contractors on such things as quality control and configuration management. It is generally quite difficult to get two contractors to produce systems and components with interchangeable parts. If they do not do so, the program office faces additional problems in spare parts procurement and logistics. Further, each added production line means an additional set of nonrecurring costs whenever there is an engineering change.

Few High-Confidence, Near-Term Benefits. The costs of competition are short-term and clear, but the benefits are long-term and uncertain. Programs can last for a decade or more. Given that the typical tenure of program managers is about three years, it is unlikely that they will be around to receive the credit for any benefits that finally accrue; consequently, they make look for strategies that return short-term benefits. In any case, they are unlikely to be rewarded merely for introducing competition; apart from exhortations in policy documents and the conventional wisdom that competition is good for everyone, few direct incentives for introducing competitive practices exist.

> The benefits of competition occur over the long term and are inherently uncertain—a disincentive for program managers.

Retrospective studies of second production source procurement programs have not been conclusive about the benefits of competition, partly because the answers depend heavily on the analytical methods used. A 1981 RAND study of the Shillelagh missile, for example, showed that analysts using the same data but different analytical procedures could produce vastly different estimates of the effect that a second production source had on procurement costs (Archibald et al., 1981). Those estimates ranged from a cost savings of 79 percent to a cost increase of 14 percent. A follow-on to the 1981 report describes five methods of estimating the cost benefits of a second production source (Birkler et al., 1990). For each method, we estimated the hypothetical single-source cost for four air-to-air missile programs (AIM-7F, AIM-7M, AIM-9L, and AIM-9M). None of the five analytical methods was unanimous in indicating that a net cost savings accrued to the government through competition in any of these programs. However, three of the five methods did show a net savings for two programs (AIM-7F and AIM-9L), and four methods did show a cost increase for one procurement (the AIM-7M).

Some uncertainty is inevitable because the various methods used for measuring savings are unavoidably judgmental. In particular, if two sources are used, one cannot know the cost that would have been incurred with a single source only. That cost must be estimated and compared with the actual cost incurred through second-source procurement. Some uncertainty about the benefits of competition is also inevitable because real cost reductions are difficult to prove and can be masked by other factors, such as changes in production quantities, requirements growth, and inflation. RAND's analysis of the Tomahawk and the Advanced Medium-Range Air-to-Air Missile (AMRAAM) programs—both of which introduced competition during the production phase—indicated cost savings were achieved (Birkler and Large, 1990). However, it was exceedingly difficult, if not impossible, to isolate and quantify a distinct cost benefit for competition from other program aspects, such as stability and maturity of the design, a firm's business base and future outlook, availability of competing systems, government and the firms' management approaches, profit levels, and overall economic environment. The path not taken is always an educated guess. When the results are very sensitive to the assumptions made, one must be cautious in drawing any conclusions.

In addition to all of the drawbacks discussed above, competition is simply seen as impractical in many cases. There may be few qualified contractors to participate in a competition, and they may not wish to compete. Contractors are sometimes deterred from entering a competition due to uncertainties about how a competition will come out and the criteria to be used in the source selection. Qualifying a second production source can also be seen as impractical because the production run is too small, the tooling for the second production line is too expensive, or the design is too complex to be transferable. In the case of subsystem components, there simply may not be enough money involved to justify the cost of funding another source.

Conditions Favorable to Competition via a Second Production Source

When is the introduction of competition a reasonable acquisition strategy? The question is not easy to answer because it is difficult to measure the future effects of competition, as suggested by the various, and sometimes conflicting, answers produced when measuring the past effects. Estimates of cost savings from competition are contingent on

- speculation about what might have happened if a second contractor had not been brought into a program
- assumptions about estimates of program cost without competition. If that estimate is too high, savings from competition or other causes would be easy to achieve. If it is too low, savings would be unlikely.

If the likelihood of breaking even on production costs is high, competition might be reasonable; we use historical data to help calculate a break-even point.

Using historical data, we can estimate the likelihood that the government would "break even" on the introduction of a competitive second source; that is, that the cost reductions would be great enough to pay for the incremental costs of introducing competition.[4] If the likelihood is high, the government might

---

[4]   Incremental costs include any nonrecurring or recurring development costs, nonrecurring investment in a manufacturing plant (facilities, general-purpose tools, general-purpose test equipment), and recurring materiel and labor costs.

reasonably elect to introduce competition in the expectation of achiev-ing other potential benefits. Likewise, a low expectation of breaking even on production cost would discourage the government from intro-ducing competition because the net dollar cost of production might increase enough to outweigh other possible benefits.

Unfortunately, no data or cost estimating relationships exist that enable one to directly estimate production costs in a competitive environment. Instead, we have historical data showing the amount by which production cost changed when competition was introduced into ongoing sole-source production programs. Multiple studies of compe-tition in procurement have been conducted over the past 30 years, with the most recent completed in the early 1990s. Those historical studies cover a wide variety of weapon systems, subsystems, and components.[5] In all of those cases, the program started with a sole-source producer, and a competitive second source was introduced later in the produc-tion run.

Whether and how much a second producer of a weapon system generates cost savings for DoD depends on the type of hardware or system that the competitors are developing and manufacturing. Our examination of the DoD's past experience with introducing competi-tion into weapon programs suggests that second producers of elec-tronics[6] have been more likely to generate savings in production costs than have second producers of missiles and ships.[7] As Table 1.2 shows, half of the DoD programs in our historical survey that involved two or more competitive producers of electronics were able to reduce overall costs by 30 percent (which turned out to be the break-even point), but

---

[5] For more detailed information about these data, see Birkler et al., 2001. Of the many sources we reviewed, we were able to obtain data from one that appears to be methodologically consistent: Birkler et al., 1990. The savings are based on actual costs or projections to the end of the program.

[6] By electronics, we are referring to those items that are mainly subsystems or small compo-nents (e.g., radios, radars, transponders, and signal converters).

[7] Our analysis treats the electronics systems and hardware separately. Most of the nonelec-tronic items are ships and missiles; there have been no instances since World War II in which aircraft were produced by competitive sources.

Table 1.2
Fraction of Programs Examined That Achieved Savings

Savings Achieved (%)	Missiles and Ships	Electronics
> 0	8/10	10/10
>10	7/10	9/10
>20	4/10	7/10
>30	2/10	5/10
>40	Nil	4/10

SOURCE: Birkler et al., 2001.

only one in ten competitive missile and ship production efforts were able to do so.

To gauge the likelihood that a government agency would recoup its costs if it were to invest in a second producer, we next applied a RAND-developed tool—the required cost reduction (RCR) methodology[8]—that determines whether competition can be reasonably introduced into the development and production phases of a variety of weapon systems. In particular, the model has allowed us to look at whether lower production costs engendered by the presence of a second producer would offset the investment in bringing that second source into the program. In recent years, we have applied this methodology to the Joint Strike Fighter, the DD(X) program (as the destroyer was then known), and to the United Kingdom's Type 45 Destroyer, as well as to numerous other acquisition programs.

A RAND-developed tool calculates the percentage of savings in production needed to offset inefficiencies with a second producer.

The RCR methodology calculates a ratio of required savings necessary to offset the additional costs introduced by multiple production sources (under competition) relative to the

---

[8] This methodology was introduced in Birkler et al., 2001. It is a modification to the original break-even analysis developed by Margolis et al., 1985.

sole-source production cost.[9] In other words, it answers the question, "How much do I need to save relative to the sole-source production costs to make competition break even?" To determine whether competition might be reasonable, we compare the RCR ratio to the values in Table 1.2. If the RCR is lower than the value for savings achieved where 50 percent of the programs saved at least that amount (i.e., there is a high likelihood of offsetting the additional costs of competition), then we view a competitive strategy as being a reasonable approach. If the percentage is higher, then competition is not a reasonable strategy to reduce costs.

The RCR factors in the cost inefficiencies that are introduced with competition. The two cost inefficiencies that are typically considered are loss-of-learning and higher investment costs.[10] Loss-of-learning occurs under a competition option because no one producer manufactures every unit. Learning (or, more formally, cost improvement) is the phenomenon where unit production costs decrease with each successive unit. However, each producer under a competitive approach will typically produce *fewer* total units over the entire production run. This reduction in the total quantity each source produces indicates that the average unit cost under competition could be higher.

Additional investment costs also occur with a competitive approach. For example, each producer must invest in tooling and manufacturing facilities to produce enough units if it wins a competition, so the overall facilities cost for a program are higher because each manufacturer needs redundant capability. Further, each manufacturer may need to engage in production design in order to produce an item (even under a build-to-print competition). Therefore, development costs may also increase under a competitive strategy.

---

[9]   This RCR ratio is defined as $RCR = (TC_1 + TC_2 - TC_{SS} + I_C - I_{SS}) / TC_{SS}$, where,
   TC_1 = recurring cost for contractor 1 under competition
   TC_2 = recurring cost for contractor 2 under competition
   TC_{SS} = recurring cost for single-source contractor
   I_C = nonrecurring cost (investment) required under competition
   I_{SS} = nonrecurring cost required under sole-source production

[10]   One could consider other influences, such as changes due to production rate efficiency and overhead. These influences are beyond the scope of this analysis.

To illustrate the potential range of answers with the RCR methodology, we built a simple spreadsheet model to evaluate competition using different key assumptions. The model calculates production costs based on three key inputs: (1) a cost-improvement slope, (2) a value for the total number of units, and (3) the nonrecurring investment cost. The model includes some simplifying assumptions:

- First, competition will equally split the quantities between the two producers (it is not a winner-take-all situation). This first assumption represents a "worst case" scenario.
- Second, the model assumes that the additional investment cost is equal to the sole-source investment cost. That is, each competing producer's nonrecurring costs are equal and do not change with production quantity.

Table 1.3 presents the results of this calculation, based on a case where the total production quantity is 1,000 units. The first column of the table represents the nonrecurring cost relative to the first production unit cost; for example, a value of 5 in the first column means that the nonrecurring cost was five times greater than the cost of the first production unit, and a value of 50 means that the nonrecurring cost was 50 times greater than the cost of the first production unit. The remaining column headings represent (in percentages) an extent to which cost improvement may occur. For example, a cost improvement of 95 percent means that unit production cost decreased by 5 percent each time total production quantity doubled; a cost improvement of 100 percent means that unit production costs remained the same when total production quantity doubled.

Table 1.3 clearly illustrates that competition is more reasonable for situations where the nonrecurring costs are relatively low and the cost improvement is minimal (flat)—for example, when nonrecurring costs are only 5 times greater than the cost of the first production unit and cost improvement is 100 percent, then the percentage of savings in production needed to offset the costs of competition is only 1 percent. Note that RCR values (i.e., the results of the calculation) over 100 percent are cases where it is impossible to achieve savings. Even if

Table 1.3
Notional RCR Values: Savings in Production Needed to Offset Inefficiencies
with Dual Sources, for 1,000 Units

Nonrecurring Costs (T1)	Cost-Improvement Slope (%)					
	100	95	90	85	80	75
1	0	5	11	18	25	34
5	1	6	12	19	28	38
10	1	7	13	21	31	43
50	5	13	23	37	56	85
100	10	21	35	56	88	137
500	50	82	132	211	340	553
1,000	100	160	254	405	655	1,073
5,000	500	777	1,224	1,956	3,176	5,237

the entire production was free, one could not offset the additional cost inefficiencies with multiple sources.

To understand when it might be reasonable to utilize competition (have dual sources), we compare the historical savings values for the ships, missiles, and electronics systems discussed in Table 1.2 with the notional RCR values in Table 1.3. We view competition as being reasonable where there is at least a 50-50 chance of achieving savings (i.e., at least half of the programs in our historical sample achieved that level of savings or more).[11] For ship and missile systems, the cases where the cost-improvement slope is greater than 90 percent and the nonrecurring costs are less than 100 times the first unit cost

Our analyses indicate that competition is more reasonable in situations when both nonrecurring costs are low and cost improvement is minimal.

---

[11]   At least half of the ship and missile programs in Table 1.2 achieved cost reductions of at least 17 percent. Similarly, at least half of the electronic systems programs in Table 1.2 achieved cost reductions of about 30 percent.

are favorable (see values shaded in green). For no cases where the cost-improvement slope was lower than 85 percent was competition seen as favorable for these weapon systems. For electronics-type systems, the range of competition-favorable values expands to include more cells (shown by the cells in green type, including those in the area shaded in green). Now, some favorable values extend to cost-improvement slopes as low as 80 percent, but the relative nonrecurring cost is still limited to values of 100 times the first unit cost and less.

Competition may also be more reasonable in situations where a greater number of units will be produced.

Table 1.4 shows a similar analysis, but where the production quantity has been cut in half (500 units). Again, a similar pattern emerges, but now fewer cells are identified as favorable. This leads to the final observation on competition: Situations with many production units are generally more favorable.

Table 1.4
Notional RCR Values: Savings in Production Needed to Offset Inefficiencies with Dual Sources, for 500 Units

Nonrecurring Costs (T1)	Cost-Improvement Slope (%)					
	100	95	90	85	80	75
1	0	5	11	18	25	34
5	1	7	13	21	29	40
10	2	8	15	24	34	48
50	10	20	33	50	75	111
100	20	35	55	83	126	189
500	100	152	229	347	530	817
1,000	200	299	448	678	1,036	1,603
5,000	1,000	1,473	2,196	3,319	5,083	7,885

Conclusions

In major defense acquisitions, the relationship between the buyer and producer is almost completely different from that assumed in the economist's model of the marketplace. While the use of competition in weapon system acquisition is widely advocated, *savings are not inevitable*. Splitting production between two contractors may in some instances result in a higher cost to the government.

As described in this paper, RAND has created a unique methodology that senior decisionmakers have used to determine if and when introducing production competition is a reasonable acquisition strategy. Over the past decade, we have developed and refined a "break-even" model, built upon previous RAND studies, that identifies how competition might be introduced into the production phases of a variety of weapon systems. Using this model, we have been able to gauge the likelihood that a government agency would recoup its costs if it were to invest in a second producer. In particular, the model has allowed us to look at whether lower production costs engendered by the presence of a second producer could offset the investment in bringing that second source into the program. In recent years, we have applied this methodology to the Joint Strike Fighter, the DD(X) program (as the destroyer was then known), and to the United Kingdom's Type 45 Destroyer, as well as to numerous other highly visible acquisition programs where the benefits of competition were being hotly debated.[12]

Because each and every acquisition is unique, our experience is that one must carefully evaluate whether introducing competition is *reasonable*. This is not an academic problem. Decisions involving billions of dollars in future procurement will be based, to some extent, on estimates of single-source versus second-source cost.

---

[12] Birkler et al., 2001; Schank et al., 2006; Birkler et al., 2002.

CHAPTER TWO

Untying Gulliver: Taking Risks to Acquire Novel Weapon Systems

John Birkler

Introduction

Traditionally, the defense acquisition system equips relatively large forces for major combat operations involving weapon systems that are produced in significant quantities and intended to be operational for decades. However, today there is a growing need to equip smaller forces to respond to asymmetrical threats using novel weapon systems that can be quickly developed and fielded.

Novel systems—often an integration of several known technologies, coupled with doctrinal and organizational changes—have more uncertainty when compared to traditional acquisition programs, and they present a challenge to the traditional acquisition process. Acquisition policies and procedures in place today are designed to deliver new systems based on a stable design to minimize risk. However, to quickly field innovative and novel systems, the acquisition community must accept precisely the uncertainties and risks that the traditional acquisition process has been deliberately designed to avoid.

The aversion to risk that is built into the current acquisition process impedes rather than encourages the development of novel systems, especially those based upon disruptive rather than evolutionary technology. Although DoD has established a number of organizations and undertaken numerous initiatives to manage the identification, test, and deployment of novel systems,[1] creating capabilities in the absence of

---

[1] Examples include the Joint Rapid Acquisition Cell, the U.S. Army's Rapid Equipping Force, the Director of Defense Research and Engineering's Rapid Reaction Technology Office,

any expressed warfighter need—that is, "technology push"—continues to run too much against the political, bureaucratic, and regulatory grain of the defense acquisition process. If it is allowed to continue, this aversion will have serious consequences for the long-term quality and capability of future U.S. combat forces. As near-term budget pressures and force modernization needs mount, spending scarce resources on capabilities that *might* become available or that *might* produce fundamental changes in mission capabilities is understandably viewed with little enthusiasm. Creating an environment that fosters innovation and novel system development is one of the tough, but fundamental, challenges facing senior leaders in DoD.

An acquisition strategy for developing novel systems cannot hinge on achieving precise cost, schedule, and performance outcomes.

This paper argues that fostering innovative systems requires a separate acquisition strategy that

- focuses on technology push and unique integrations of existing and emerging technologies;
- emphasizes flexibility, including an overt willingness to accept risks;
- allows easy and quick termination of programs not yielding expected benefits;
- enables early test and demonstration of military utility.

In other words, we argue that fostering innovative systems requires a strategy that is more streamlined and less tied to achieving precise estimated cost, schedule, and performance outcomes in order to provide improved or unique capabilities to the warfighter as quickly as possible.

---

the Defense Advanced Research Projects Agency, the U.S. Air Force's Quick Reaction Cell, the now-closed Air Force battle labs, Big Safari, et cetera.

In this paper, we will first define what we mean by novel systems and describe their special features, and then we will outline the necessary elements of a strategy for developing these systems.

What Is a Novel System?

Novel systems differ from legacy or conventional systems on several dimensions: design, operational employment, outcomes, production run, and operational life. Figure 2.1 compares novel systems to conventional systems in terms of these dimensions (the descriptive adjectives used are deliberately simplified in order to emphasize the extent of the differences).

Conventional systems—for example, the Joint Strike Fighter, the F-22 fighter, and the Navy's new destroyer, DD(X)—may contain the latest technologies, but this alone does not constitute a "novel system." We know how to classify conventional systems (e.g., tactical aircraft, surface combatant); we know how we are going to use them; they fill an official capability gap; and we know about how many we plan to buy.

On the other hand, in novel systems—such as the F-117 Stealth Fighter, novel mine-counter measures, and robotic ground vehicles— the following factors are less certain:

Figure 2.1
Comparison of Conventional and Novel Systems

	DIMENSIONS	
Conventional Systems		_Novel Systems_
Follow-on	Design	New
Evolutionary	Technology	Disruptive
Established	Operational employment	In formulation
Predictable	Outcomes	Uncertain
Large	Production run	Uncertain
Long	Operational life	Uncertain

- The design of a novel system is new in overall concept or in integration of existing/use of emerging technologies, or both, so that the development outcomes (mainly performance and cost) cannot be confidently predicted on the basis of studies alone.
- The operational employment doctrine has not been clearly defined and demonstrated and is therefore subject to substantial uncertainties and change.
- The eventual size of the production run and the subsequent operational life are uncertain (an obvious consequence of the uncertainties surrounding the cost, capabilities, and operational concept of the system).
- The nature of the key uncertainties is such that they can be resolved only through development and test of a system or through prototypes, hopefully at a cost that is commensurate with the potential value of the system.

The case of unmanned aerial systems (UASs) during the 1980s and early 1990s illustrates many of these characteristics. It had been technically possible to build generic UAS platforms for several decades, and many had been built and used as aerial targets and reconnaissance drones. Throughout that period, various combat and combat-support roles had been posited for UASs, but every proposed application raised a host of troublesome issues: exactly how would UASs be controlled, especially in situations demanding deviation from the original mission plan? How would the information normally obtained visually by the pilot be acquired and translated into mission-relevant decisions? How would safety be ensured during peacetime operations over populated areas? And so on. Despite many studies and a small number of actual development projects that were canceled early, few UAS programs were actually completed in the United States during this time. How can we explain this? A 1997 study (Sommer et al., 1997) suggested a range of possible factors.

Novel systems involve uncertainty not only with regard to design, but also in terms of operational employment and possible outcomes.

The cause of the poor track record of UAV[2] programs in the United States is not entirely clear. Certainly, the mere fact of their being *unmanned* vehicles cannot be the cause. After all, the United States has had great success with other unmanned systems, ranging from interplanetary spacecraft and satellites to cruise missiles and submersibles. What, then, makes UAVs unique? A possible explanation is that UAVs in general have never had the degree of operational user support necessary to allow their procurement in sufficient quantities (perhaps because of funding competitions from incumbent programs, or because of the conjectural nature of their capabilities). Thus, the learning curve is never ascended, multiple failures occur, risk tolerance decreases, unit costs rise as a result, and user support decreases yet further in a diminishing spiral.

In addition to these factors, the acquisition process itself (as defined in DoD's 5000 series of directives) is simply not congenial to programs with a range of important uncertainties. A key problem in developing novel systems lies in the sequence of decisions and actions involved in the defense acquisition process. Early on, when DoD is trying to decide how to fill a capability gap, a series of studies are performed, followed by a Milestone A decision, at which point the system concept to be developed is clearly defined and the sponsor commits to funding for development and initial production. In the case of novel systems, major uncertainties and risks are not likely to be adequately resolved at this milestone; thus, the acquisition process eliminates them from further consideration. The funding needed to explore and resolve the major uncertainties generally exceeds that which could be obtained by a project, unless it is directly coupled with a major acquisition program. Thus, uncertainties are unresolved, and progress is stifled.

> **Major uncertainties in novel systems generally will not be resolved by Milestone A; this means they are eliminated from further consideration.**

---

[2]   Unmanned aerial systems (UASs) were formerly called unmanned aerial vehicles (UAVs).

In the commercial world, some of the most successful products are ones consumers never knew they needed until they came on the market—iPods, cell phones, digital cameras, Blackberries, GPS navigation systems, Bluetooth headsets, et cetera. To achieve these breakthroughs, businesses accept a greater amount of risk in developing some of their product lines.

A Strategy for Fielding Novel Systems Concepts

The characteristics of novel systems are so different from those of the systems for which the present acquisition process was designed that we believe "tinkering" with the present process is impractical. To formulate more appropriate procedures, we identify major elements of an acquisition strategy that would be more consistent with the special features of novel systems and with the expected environment of urgency that might attend their development.

A new strategy is needed to oversee development and demonstration of novel systems.

Provide an Environment That Fosters New Concepts for Systems and New Concepts of Operations. To provide a rich source of new options to address emerging threats in a timely manner, DoD needs (1) staff who combine both technical and operational experience and skills and (2) a culture in which innovation is constrained only by perceptions of technical feasibility and relative operational value compared to other innovative investment ideas, not by current doctrine on force composition and employment.

Monitor Civilian Technologies That Could Be Integrated in Unique Ways to Give Warfighters New Capabilities.[3] This second element is

---

[3] The idea of a single technological breakthrough, while popular, is belied by fact. Advances have not come with the introduction of a spectacular new technology, but with the integration of several known and often rather mundane existing technologies.

needed because the level and scope of the private sector's investment in fast-moving commercial technologies outpace DoD's efforts to shape the marketplace. Integrating and morphing commercial technologies into unique and new warfighting capabilities should be opportunity driven, rather than need driven, as is currently the case.

Upon Successful Demonstration of a New System, Permit Early, Provisional Fielding and Operation Before Completion of Full Maturation Development and Associated Testing. The third element of the strategy focuses on the later phases of acquisition. Today's acquisition procedures demand (1) extensive effort toward system maturation to minimize support costs, together with (2) extensive operational testing to ensure that no significant lingering problems and deficiencies exist. The interrelationships of funding, testing, and buying hardware that occur at this point should be examined with the objective of relaxing their interdependency. Delinking funding approvals from testing status could allow a novel system to proceed through an appropriation threshold and would break the link with service "requirements." This delinking will reduce costs and project duration, getting the system to the warfighter sooner, but with less maturity than traditional systems.

> Our most radical proposal—and most powerful tactic to acquire novel systems—is to delink funding, testing, and buying.

This element is the most radical one of the acquisition strategy recommended in this paper, and it poses the most challenging implementation problems, but it potentially contains the most powerful tactic for moving an innovative new system concept to early operational capability. To implement this element will require establishment of "experimental" operational units designed to receive and operate systems that are not quite technically mature, that are not fully provisioned with support and training aids, and that lie outside the main thrust of acquisition policy for traditional major defense acquisition programs.

Encourage Timely and Visionary Decisions on Novel Programs by Decreasing the Need for Extensive Staff Support and Documentation

and by Giving a Few Junior Officials More Authority for These Programs. The fourth element of the strategy is needed to remove the extensive and lengthy documentation and review procedures now required for milestone approval, especially at Milestone A. Those procedures were designed to ensure a full examination of all alternative concepts, to create a broad-based consensus on the selected concept and to manage risk. Such extensive documentation and reviews might make sense when the proposed new system concept is an extension of previous design concepts and operational employment strategies because a broad accumulation of experience and historical data are available. However, such accumulation of experience and corresponding data do not exist for novel systems concepts. Further, DoD may have put too much faith in senior officials to make judgments and decisions about concepts with which they and their staffs have little or no experience. Those officials operate in an environment that severely criticizes them for any unsuccessful project. Many of our current "rules" are designed to govern the perceived excesses of acquisition officials. Giving junior executing officers more flexibility and responsibility, while holding them accountable, may be a more realistic and more effective approach.

> DoD will need to establish "experimental" operational units designed to receive and operate systems that are not quite technically mature.

Systematically Accumulate Lessons Learned About Managing the Development and Demonstration of Novel Systems and Operational Concepts. Almost without exception, novel systems have been conceived, designed, and developed outside of the conventional acquisition system. All of the services have a quick reaction capability that functions outside normal acquisition rules and focuses on very quickly fielding innovative concepts and novel systems. Similarly, DoD's Defense Advanced Research Projects Agency and, in the private sector, Lockheed Martin's Skunk Works and Boeing's Phantom Works have acquired extensive experience with novel systems.

The point is that there are people and organizations that have significant experience with technology development, cleverly integrating existing technologies and making changes in tactics that can help DoD more effectively bring innovation and novel systems to warfighting. An in-depth, systematic analysis of successful as well as ineffective organizational and managerial attributes for fostering innovation in the absence of a competitive, free market would help DoD establish new organizations with explicit charters for experimentation, testing and learning, and demonstration.

Conclusions

Each of these elements has been applied in the past under special circumstances, with beneficial results—for example, in developing the F-117, sea-launched ballistic missiles, cruise missiles, and UASs. But "special case" applications hinder the systematic development of management expertise and effective management processes. Moreover, these applications were difficult to implement and transition to operational forces due to an acquisition environment that favors more detail about the end stages of a program than these mechanisms can provide.

Hence, we believe it appropriate and desirable to devise a less formal, less "standard" path for the acquisition of novel systems, based on the strategy outlined above. By understanding the environmental attributes to foster novel systems, DoD will signal that novel system ideas and concepts will be encouraged, and it will bestow a military advantage to U.S. warfighters as the first nation to develop and use them.

CHAPTER THREE

Dollar Value and Risk Levels: Changing How Weapon System Programs Are Managed

Robert Murphy and John Birkler

Introduction

Currently, acquisition programs are grouped and then managed at the Office of the Secretary of Defense (OSD) by dollar value; depending on the dollar value, OSD provides different levels of oversight and different management processes. This approach has been constantly refined over the years without having produced any noticeable improvement in terms of reducing the cost growth, schedule slippage, and performance shortfalls that continue to plague the acquisition of weapon system programs. This paper argues for a different paradigm: The level of overall risk inherent in a program should be the main basis for determining the process and level of review a project should receive.[1]

Drawing upon examples from warship acquisition programs, this paper also argues that inadequate assessment and management of various discrete program risks result in adverse cost, schedule, and performance outcomes. We examine existing scales for assessing some of these discrete program risks and make recommendations to better assess and manage several programs within the Defense Acquisition Management System.

---

[1] Cost is a factor that must be considered when determining the level of review. A multi-billion dollar program requires high-level review because even a small amount of cost growth involves large dollar amounts.

Managing by Risk Level Versus Dollar Value

Currently, OSD requires review of acquisition programs and decisions by senior officials on the basis of a program's dollar value, irrespective of risk, as shown in Table 3.1.

OSD oversight is based on a program's dollar value, irrespective of risk.

However, some very costly projects might have significantly less risk than projects of similar cost, and thus should require less oversight as well as the use of different criteria at milestone reviews.[2] Conversely, projects may cost little but have a lot of risk because they tend to push the state of the art in technology and may also involve novel business or design processes that may require more comprehensive oversight than just dollar value would otherwise indicate. An excellent example of this type of program—the Advanced SEAL Delivery System (ASDS)—was discussed in a May 2007 report by the U.S. Government Accountability Office (GAO). The ASDS is a Special Operations Forces' battery-powered submersible carried to a deployment area by a submarine. The operating parameters for the submersible required development of batteries that would push the state of the art in that technology. The initial design, construct, and deliver contract was awarded for $70 million in 1994 for delivery in 1997; because of the dollar value, Milestone Decision Authority (MDA) resided with the Navy, which ultimately accepted delivery of ASDS in 2003 in "as is" condition at a cost in excess of $340 million. GAO concluded that "had the original business case for ASDS been properly assessed as an under-resourced, concurrent technology, design, and construction effort led by an inexperienced contractor,

---

[2] For example, the Navy is about to restart construction of two DDG 51-class destroyers at a cost in excess of several billion dollars. Over 60 destroyers of this class have already been delivered or are in the final stages of construction. Because of this track record, restarting construction of two new DDG 51s will no doubt expose the Navy to a far less risk of adverse cost, schedule, and performance outcomes than construction of three multibillion DDG 1000-class ships, which are now being designed and just entering construction.

Table 3.1
Basis and Level of Program Oversight

Program Acquisition Category (ACAT)	Basis for ACAT Designation	Milestone Decision Authority (MDA)
I	Estimated total expenditure for research, development, test, and evaluation (RDT&E) of more than $365 million or for procurement of more than $2.190 billion	ACAT ID: Under Secretary of Defense for Acquisition, Technology, and Logistics ACAT IC: Head of DoD Component (e.g., Secretary of the Navy) or, if delegated, DoD component acquisition executive (e.g., Assistant Secretary of the Navy for Research, Development, and Acquisition)
II	Estimated total expenditure for RDT&E of more than $140 million or for procurement of more than $660 million	DoD component acquisition executive or designate (e.g., program executive officer)
III	Does not meet criteria for ACAT II or above; less than a MAIS program	Designated by DoD component acquisition executive at the lowest level appropriate (e.g., program manager)

SOURCE: DoD, December 8, 2008.
NOTE: Estimated expenditures are in FY 2000 constant dollars.

DoD might have adopted an alternative solution or strategy" (GAO, May 2007, p. 13).

Focusing on Causes Rather Than Consequences

Risk, or the exposure to the chance of failure, is a word heard frequently in the acquisition community. All acquisition programs face risk of some form or another. Arguably, any new major weapon system that could be developed, produced, and fielded with no risk involved is probably not worth acquiring.

Overtly or otherwise, much of a program manager's time is spent managing risk. After all, the Defense Acquisition Management System, shown in Figure 3.1 is, in essence, a risk-management process designed to ensure success in the timely delivery of weapon systems that meet warfighter requirements while staying within budget.

The risks most frequently mentioned by defense acquisition officials are cost growth, schedule slippage, and performance shortfalls. This is not surprising as cost, schedule, and performance are the primary attributes by which programs are assessed for success or failure. Moreover, the Defense Acquisition University (p. 2) teaches that cost, schedule, and performance are the risk factors that program managers must assess and manage "to ensure that DoD is acquiring optimum systems that meet all capability needs."

> **Managing by cost, schedule, or performance risks is reactive; managing by discrete programmatic risks is more proactive.**

Assessment of cost, schedule, and performance is clearly a management task, and a good program manager assesses these risks using periodic data accumulated into management reports to identify problems, regain lost ground, and then stay on track. However, these are broad measures of risk. A better program manager proactively manages by using discrete program risks, submeasures that allow him or her

Figure 3.1
The Defense Acquisition Management System

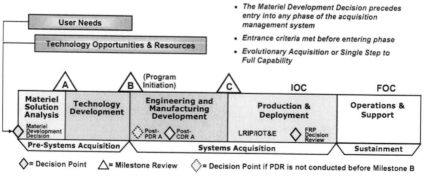

SOURCE: DoD, December 2, 2008.

to look ahead and act to avoid adverse cost, schedule, and/or performance trends and outcomes. In other words, managing by cost, schedule, and performance measures is akin to driving a car while looking solely in the rearview mirror: it is possible, but only if the road stays straight. A better driver looks mostly out the windshield, with only an occasional look in the mirror; this driver anticipates and easily handles curves in the road.

In this paper, we focus on five discrete programmatic risk categories:

- technical
- system integration
- design
- production
- business.

Taken together, these risk categories portray overall acquisition program risk.[3] They interact in numerous ways to affect a project's cost, schedule, and/or performance outcomes: Obviously, technologies that do not work affect performance, but so can poor business decisions that increase cost and lead to features being deleted from the weapon system to remain within budget.

The Defense Acquisition Management System appears to adequately recognize that incorporation of new technologies into a weapon system presents risk, providing metrics to systematically assess this type of risk. Time is also provided in the acquisition process for system integration matters to be identified and resolved, although there is room for improvement. However, as will be discussed in subsequent examples, new approaches in design, production, and business areas of acquisition programs do not appear to receive the same skepticism and comprehensive oversight that new technologies and systems receive.

---

[3] For simplicity, risks involved in fielding, operating, and maintaining the weapon system are not addressed in this paper.

Well-Defined Process for Assessing Technical Risk Is in Place

"Technical risk" is exposure to the chance that development of critical technologies will not meet program objectives within cost and schedule constraints. In assessing technical risk, program managers must address the uncertainty in their estimates about how much time and effort it will take to make new technologies work. The importance of technical risk is well understood in the acquisition community. For example, DoD guidance states that "the management and mitigation of technology risk . . . is a crucial part of overall program management and is especially relevant to meeting cost and schedule goals" (DoD, 2008, para. 3.7.2.2).

> **The Defense Acquisition System Framework incorporates assessment of technical risk and system integration risk but puts less emphasis on other types of risk.**

Technical risk is also extensively addressed in the Defense Acquisition Management System. The system recognizes evolutionary acquisition as the preferred DoD strategy for rapid acquisition of mature technology for the user. One purpose of evolutionary acquisition (i.e., delivering capability in increments through spiral or incremental development) is to provide time to better manage technology risk and avoid adverse cost and schedule outcomes that often result from trying to achieve difficult requirements in one step.

DoD has also established a well-defined process based on Technology Readiness Levels (TRLs) to categorize technical risk and help ensure that key decisionmakers understand the risk of incorporating different technologies into weapon system acquisition programs (the TRLs are described in Table 3.2). Using this process, program offices conduct a technology readiness assessment under the auspices of the DoD Component Science and Technology (S&T) Executive; the Deputy Under Secretary of Defense (S&T) evaluates the technology readiness assessment and forwards findings to the Overarching Integrated Product Team leader and Defense Acquisition Board.

The TRLs are a good proxy measurement for technical risk: The lower the readiness level, the more development needed to incorporate the technology into a weapon system; and the more development

Table 3.2
Technology Readiness Levels

Technology Readiness Levels
1. Basic principles observed and reported
2. Technology concept and/or application formulated
3. Analytical and experimental critical function and/or characteristic proof of concept
4. Component and/or breadboard validation in laboratory environment
5. Component and/or breadboard validation in relevant environment
6. System/subsystem model or prototype demonstration in a relevant environment
7. System prototype demonstration in an operational environment
8. Actual system completed and qualified through test and demonstration
9. Actual system proven through successful mission operations

SOURCE: DoD, May 2005.
NOTE: See Mankins, 1995.

needed, the greater the risk. Overall, technology risk has been handled fairly well in warship acquisition programs, which tend not to push the state of the art in technology as far as do weapons and sensors. A recent example is the USS *Virginia*, which incorporates various new technologies[4] and was still delivered within four months of the original schedule established a decade earlier (Casey, 2007).

System Integration Risk Is Assessed, but at Later Stages

The acquisition community also assesses system integration risk, but it lacks effective tools to measure and categorize this risk early in a program's life cycle. "System integration risk" is exposure to the chance that new and existing technologies being employed in a weapon system may not work together and/or interact with operators and maintainers to meet program objectives within cost and schedule constraints. System

---

[4] For example, a nonpenetrating photonics mast versus a periscope, a DC electric system, Lightweight Wide Aperture Sonar Arrays, et cetera.

integration can be an issue within an individual acquisition program (e.g., when subsystems fail to interact). It can also be an issue between acquisition programs: Many programs develop capabilities that are a component of a larger warfighting capability; individually, the component programs might appear to be a low or moderate risk, but in combination with other programs, the overall risk might be much higher due to coordination and integration issues. A classic example occurred during the Grenada invasion when Army and Navy communications systems did not interact well during the joint operation.

System integration risk is extensively treated after Milestone B, during the engineering and manufacturing development (EMD) phase, at which time a program should demonstrate system integration, interoperability, safety, and utility (DoD, 2008, para. 3.7.1.1). While appropriate attention is given to system integration risk during this phase, this assessment occurs after the second of three milestones in the process, when programs have typically built up so much momentum that they are difficult to stop, regardless of performance or progress. Early consideration of system integration risk—at Milestone A—by senior decisionmakers could result in developing and funding integration-risk mitigation plans that could considerably improve acquisition outcomes.

Combat systems in warships provide an example of the problems that arise when decisions are made without adequate consideration of system integration risk.[5] For example, early decisions on systems architecture and processing approaches made without adequate consideration of risk led to cost, schedule, and performance problems with submarine combat systems for the SSN 688I, SEAWOLF, and Australian Collins-class submarines. According to a report for the Parliament of Australia discussing the Collins-class submarine (Woolner, 2001),

> Of the early decisions in the Collins program, the one which was to have the most public effect was that concerning the nature of the

---

[5]   A combat system integrates information from sensors, synthesizes this information for combat commanders, and provides fire control solutions and guidance to weapons.

vessels' Combat Data System (CDS). It has been the subsequent failure of this system to meet its design requirements that has left the submarines with a severely impaired combat capability.

By the end of 1982, . . . [the Royal Australian Navy (RAN)] had decided that the electronic combat systems of the new boats would be fully integrated. Instead of the then standard central computer performing all data analysis, the new submarine CDS would use a data bus to distribute information to a number of smaller computer work stations.

The report then goes on to discuss the lack of appreciation for the risk of switching to the new integrated architecture for combat systems.

The RAN was not alone in its 'grand folly'. . . . The Australian information technology (IT) industry assured the RAN of both the feasibility and inherent advantages of a fully integrated combat system and of its ability to contribute to such a program.

Moreover, the RAN was not the only navy to think that the future of combat data processing lay with fully integrated systems. The USN [U.S. Navy] specified the same concept for its [BSY-2] Integrated Combat System for the U.S. Navy's Seawolf class nuclear attack submarines. This was an even more costly failure than the Collins CDS, absorbing . . . $1.5 billion [in U.S. dollars] before it was cancelled.[6]

Tools for assessing system-integration maturity earlier on have been proposed. For example, Sauser et al. (2008) have proposed a System Readiness Level (SRL) index that would incorporate the current TRL scale as well as an Integration Readiness Level (IRL) scale. The IRL scale they describe would use nine levels, which appear compatible with the widely used TRLs and appear to be a good proxy

---

[6]   The original citation mistakenly attributed this to the BSY-1 program.

measurement of system integration risk. The proposed IRLs are listed in Table 3.3.

The Risks of Design Process Management Are Not Well Understood

"Design risk" is exposure to the chance that the weapon system's design will not result in effective operation or be easy to produce. It is axiomatic that a good design is essential to a weapon system's performance, but the impact of design on a weapon system's production cost and schedule outcome is not as well appreciated. However, decisions made early in the design process quickly establish not only the performance but also the ease of manufacture and resultant cost of the weapon system. While the ability of the design to operate effectively can be considered a subset of technical risk, a more holistic approach

Table 3.3
Integration Readiness Levels

Integration Readiness Levels
1. An **interface** between technologies is identified with sufficient detail to allow characterization of the relationship.
2. There is some level of specificity to characterize the **interaction** (i.e., ability to influence) between technologies through their interface.
3. There is **compatibility** (i.e., a common language) between technologies to orderly and efficiently integrate and interact.
4. There is sufficient detail in the **quality and assurance** of the integration between technologies.
5. There is sufficient **control** between technologies necessary to establish, manage, and terminate the integration.
6. The integrating technologies can **accept, translate, and structure information** for their intended application.
7. The integration of technologies is **verified and validated** with sufficient detail to be actionable.
8. Actual integration is completed and **mission qualified** through test and demonstration in the system environment.
9. Integration is **mission proven** through successful mission operations.

SOURCE: Sauser et al., 2008.

is for a program manager to assess the chance that the design process to be employed for the weapon system will generate an effective, easy-to-produce weapon.

The design process necessary for an effective and producible weapon system involves complex interactions between designers, suppliers, production experts, planners, and estimators. Design process complexity has also increased with the availability of more sophisticated design tools such as electronic product models and computational techniques (e.g., finite element analysis).

Outcomes from two current acquisition programs—the United Kingdom's ASTUTE-class submarine and the U.S. Navy's LPD 17-class of amphibious transport dock ships—demonstrate why senior decision-makers in the OSD acquisition process need to better understand the risks new design processes and tools present. The ASTUTE was the first UK submarine to be designed through use of an electronic, three-dimensional computer product model. The prime contractor's inability to manage this new process resulted in extensive delays when design products needed to build the ship were late. General Dynamics ultimately had to be hired to augment and manage the final stages of the submarine's detail design process. Because of design and other problems, the ASTUTE program has overrun cost greatly and is years behind schedule.

Two current acquisition programs illustrate why better insight is needed into the risks of new design processes and tools.

With LPD 17, the U.S. Navy competed the design and production of the first three ships of the class using as major evaluation and award criteria (1) the plans for accomplishing detail design and other functions, (2) Integrated Product Data Environment (IPDE) tools, and (3) life-cycle cost projections; these criteria were given more weight than price (Comptroller General of the United States, 1997). The then Avondale Shipyard in New Orleans, Louisiana, partnered with a firm that was developing a new ship design IPDE tool and won the competition. Subsequently, the LPD 17 experienced considerable cost growth (about 70 percent) and schedule delays (Congressional Research

Service, 2008, p. 12). GAO attributed much of this cost growth to the new design tool (GAO, July 2007):

> In the LPD 17 program, the Navy's reliance on an immature design tool led to problems that affected all aspects of the lead ship's design. Without a stable design, work was often delayed from early in the building cycle to later, during integration of the hull. Shipbuilders stated that doing the work at this stage could cost up to five times the original cost. The lead ship in the LPD class was delivered to the warfighter incomplete and with numerous mechanical failures.

Senior decisionmakers should require a program manager proposing to use new design processes, tools, or organizations to design a weapon system to justify selection of the new process, tool, or organization and develop an appropriate risk mitigation plan. An example of a design process mitigation plan comes from the VIRGINIA-class submarine program. Prior to VIRGINIA-class construction using a new Integrated Product and Process Development (IPPD) approach, Electric Boat started

> a representative section of the ship about a year early with a portion of that section started about two years early. This early, controlled, closely monitored ship construction effort ensured thorough preparation for full-ship application and high confidence in the new process. (General Dynamics Electric Boat, 2002, p. 33)

Evaluation of Production Risks Lacks Rigor

An earlier and more rigorous evaluation of production risks could save DoD much difficulty and taxpayers a lot of money. "Production risk" is exposure to the chance that the facility, labor, manufacturing processes, and procedures will fail to produce the weapon system within time and cost constraints. Producibility—or "production capability"—is a function of the design; production facilities; management skills, processes, and experience; and workforce skills and experience. DoD requires assessment of contractors' production capability before production contract award in the production and deployment phase, but this may be too late because, at this point, production may be locked in by the organization that won the design contract. Moreover, in the authors' experience and as exemplified in the LPD 17 source-selection criteria discussed earlier, the production category of risk does not receive the same emphasis in selecting a shipbuilder as other factors, such as design concepts, past performance, and estimated cost.

The Navy's DD 963-class of destroyers and LHA 1-class of amphibious assault ships are classic examples of programs in which DoD considered design and production risk acceptable when awarding contracts, but which went on to experience about the worst of every production factor possible. These ships presented little technical and system integration risk, but ended up far behind schedule and over cost due in part to identifiable production risks. Contracts were awarded to the lowest bidder, Litton Industries, which owned the Ingalls shipyard in Pascagoula, Mississippi. In the late 1960s, Litton Industries decided to invest in an expansion of design and production facilities for warships, building a new shipyard on the west bank of the Pascagoula River, across from its existing shipyard. The new shipyard was designed to be operated with a new production control system using modular techniques for building ships (Northrup Grumman, 2008).

Acquisition programs can end up far behind schedule and over cost when production risks are not adequately assessed.

After the award of the LHA- and DD 963-class contracts to Ingalls for nine LHAs and 30 DD 963s in the late 1960s, Ingalls' management decided to shift construction of some commercial container ships from the old, conventional yard to the new facility (Northrup Grumman, 2008). The expectation was that doing so would allow the new facility to start up and have any problems worked out while the LHA and DD 963 were being designed. However, production of the container ships using the new control system led to delays; consequently, the ships were occupying facilities and using manpower needed to start production of the LHAs and DD 963s. Production of the LHAs and DD

963s fell far behind and, in combination with other problems (design-related issues, inflation, et cetera), the costs were overrun substantially and the ships were late (GlobalSecurity.org, 2008).

A greater emphasis on evaluating production risks could have saved an enormous amount of time and money, but the promised cost savings resulting from construction in a new state-of-the-art ship fabrication and assembly facility proved too good to be true. The assessment that the facility would be derisked by building container ships first turned out to be wrong, and meanwhile, two entire classes of ships had been priced and placed under contract.

A promising approach, initiated by the Missile Defense Agency, may provide program offices across DoD with better insight about production risk. The agency extended the notion of TRLs to engineering and manufacturing by developing Engineering and Manufacturing Readiness Levels (EMRLs) to assess the maturity of a program's design, related materials, tooling, test equipment, manufacturing, quality, and reliability levels. There are five EMRLs, as shown in Table 3.4.

Table 3.4
Engineering and Manufacturing Readiness Levels

Engineering and Manufacturing Readiness Levels
1. System, component, or item validation in laboratory environment or initial relevant engineering application or breadboard, brass-board development.
2. System or components in prototype demonstration beyond breadboard, brass-board development.
3. System, component, or item in advanced development. Ready for low-rate initial production.
4. Similar system, component, or item previously produced or in production. System, component, or item in low-rate initial production. Ready for full-rate production.
5. Identical system, component, or item previously produced or in production. System, component, or item in full-rate production.

SOURCE: DoD, 2005.

The Risk of Early Business Decisions Is Not Fully Appreciated

Business decisions made early in a program's life can significantly affect cost, schedule, and performance outcomes. "Business risk" is exposure to the chance that the overall acquisition strategy for a program will not result in the desired cost, schedule, and/or performance outcomes. Decisions about the process to select who will build the weapon system, the standards to which it will be built, and the schedules for designing and building it all entail risk that is not always appreciated up front. To evaluate business risk, program managers should assess the following: (1) the extent to which the acquisition strategy can result in selection of the most effective, efficient design and most effective, efficient production entities; (2) whether cost estimates and schedules are valid; (3) whether proper government oversight organizations are in place; and (4) whether project personnel with proper training and experience are available.

A good example of early business decisions gone bad is the Navy's Littoral Combat Ship (LCS) Program. The lead ship, USS *Freedom* (LCS 1), was recently delivered after experiencing substantial cost overruns and delivery delays. In congressional testimony given to explain these outcomes, the U.S. Navy (2007) identified the following tenets of the new business model used to acquire the LCS:

- Construction of LCS seaframes in midtier shipyards that "perform predominately commercial work, maintaining business processes and overhead structures that keep them competitive in the world market" (i.e., little warship experience).[7]
- "A rapid 24-month build cycle for each seaframe, as opposed to the five or more years that have become the norm in naval shipbuilding."

---

[7]   To better understand the differences between military and commercial shipbuilding, see Birkler et al., *Differences Between Military and Commercial Shipbuilding: Implications for the United Kingdom's Ministry of Defence*, Santa Monica, Calif: RAND Corporation, MG-236-MOD, 2005.

- "The LM lead ship detail design and construction effort was initiated simultaneously and the lead ship commenced construction only seven months after the start of final design (i.e., concurrent design/build)."
- "In order to address the challenges of technical authority under this environment (reduction in NAVSEA technical personnel), in February 2003, NAVSEA and PEO Ships made two joint decisions. The first was to work with the American Bureau of Shipping (ABS) to develop a set of standards (Naval Vessel Rules) that could be applied to non-nuclear naval combatant ships. The second was to utilize ABS to class[8] both LCS and DDG 1000 using the new rules."

No doubt there were good arguments for these individual program tenets. However, the cumulative effect of the risks involved in building a new design warship in small commercial shipyards with little warship experience during a rapid, concurrent design/build process and to a set of technical standards themselves under development appears to have been greatly underappreciated. In that same congressional testimony, the Navy identified cost drivers for LCS 1 as "concurrent design-and-build while incorporating Naval Vessel Rules (NVR), reduction gear delays created by a manufacturing error, and insufficient program oversight" (U.S. Navy, 2007). The risks inherent in utilizing an entirely new business model to acquire warships were obviously neither adequately assessed nor managed.

One way to avoid such risk would be to require program managers proposing new and/or radical business models to fully justify why the new model is superior to past practice, recommend more frequent assessment points than now required by the Defense Acquisition Management System, and incorporate exit strategies in contracts for the government to use if the program fails to meet expectations.

---

[8] The American Bureau of Ships is known in the commercial shipping industry as a "classification society," which is an organization that sets standards for design and construction of vessels and integral machinery within.

Conclusions

The Defense Acquisition System Framework has sufficient tools and allows time for proper assessment and management of technical risk and, to some extent, of system integration risk. However, design, production, and business risks are not always adequately assessed and managed. As shown in this discussion, scales exist that represent good proxy measurements of technical, systems integration, engineering, and production risks; what is missing are descriptive levels that could be used to assess and categorize design and business process risk. We recommend that DoD explore establishing such levels and, in Tables 3.5 and 3.6, offer starting points for doing so (based on the authors' experience), which may help program managers more carefully consider these risks.

In addition, we recommend the following actions to better assess and manage program risk overall:

- Assess, categorize, and individually review each technical, system, design, production, and business risk of a program at each milestone in the Defense Acquisition Management Framework.
- Require program managers to justify new or radical approaches to design, production, or business processes and develop and implement risk mitigation plans and/or contract off-ramps.

Table 3.5
Proposed Design Process Levels

Design Processes
1. New, unproven processes. New design tools under development. New design organization.
2. Large expansion of existing design organization. Many new designers and supervisors unfamiliar with design tools and processes.
3. Existing design organization using radically changed design tools, processes, and/or technologies.
4. Experienced design organization using new design tools with proven processes.
5. Experienced design organization using existing, proven design tools and processes.

Table 3.6
Proposed Business Process Levels

Business Processes
1. Using a new, unproven approach to source selection. Encouraging new sources of supply. Acquiring new technologies without well-established cost-estimating relationships. Requiring new government and/or contractor organizations to be formed.
2. Using new procurement process in established industry. Cost-estimating relationships only at high levels. Requires expansion of government and contractor organizations.
3. Evolutionary change from prior acquisition strategies. Good cost-estimating relationships. Existing government and contractor organizations can easily adapt to changes.
4. Using same approach to buying similar products. Well-established cost-estimating relationships exist. Experienced government and contractor organizations involved.
5. Acquiring more of what has been successfully bought before. Using the same contractor and government organizations.

Although such tools would enhance the ability of program offices to assess and manage risk, DoD should also consider changes in oversight. As stated at the outset of this paper, the current acquisition system requires review and decisions by senior officials on the basis of a program's dollar value, irrespective of risk. A better use of their limited time may be to focus on programs with high risks, letting less-senior officials deal with lower risk programs, regardless of dollar value. For example, DoD could

- lower the MDA level for future milestones down
  —two levels for programs with low risk in all risk categories[9]
  —one level for programs with moderate risk in all risk categories[10]

---

[9]   Determination of what constitutes "low risk" is obviously subjective. For our purposes, "low risk" would be technology and integration levels 8 and 9 and EMRL, design, and business levels 4 and 5.

[10]   For our purposes, "moderate risk" would be TRL and IRL 5 and 6 and EMRL design and business levels 3.

- continue to follow the patterns for decision authority as established in the Defense Acquisition Management System for any program with greater than moderate risk in any of the five categories of program risk.

In this way, senior decisionmakers might be able to better concentrate their limited time on the real potential problem areas in a program before problems occur, and direct actions to be taken to avoid and/or mitigate potential problems.

CHAPTER FOUR

Improving Acquisition Outcomes: Organizational and Management Issues

Irv Blickstein and Charles Nemfakos

Introduction

Despite years of change and reform, DoD continues to develop and acquire weapon systems that it cannot afford and cannot deliver on schedule. The U.S. Government Accountability Office reported, for example, that research and development costs in 2008 for selected major programs were 42 percent higher than originally estimated and that the average delay in delivering initial capabilities had increased to 22 months (March 2009, p. 1). Also, William Lynn, the new deputy secretary of defense, stated that it will be very difficult to sustain a force large enough to meet demands if current acquisition trends continue. This suggests that the new administration will have to find ways to halt traditional cost growth associated with fielding new capabilities—or consider program terminations (Lynn, 2009, p. 16).

Many of the problems that contribute to poor cost and schedule outcomes are systemic to the way that the acquisition process is organized and managed in DoD. It is our purpose in this paper to discuss a few of these problems and how they may be contributing to inefficiency and unrealistic expectations. These issues include

- the service chiefs' role in the acquisition process
- the combatant commands' (COCOMs') role in the requirements process
- the impact of joint duty on the acquisition process
- a growing emphasis on management processes at the expense of workforce initiative.

The Service Chiefs' Role in the Acquisition Process Is Too Limited

Stronger ties are needed between service leaders who understand warfighting needs based on operational experience and those who are responsible for acquiring weapon systems to meet those needs. Over time, a schism has developed in some of the military departments between the service chiefs, who validate warfighting requirements, and the service acquisition chiefs who develop and acquire new weapon and information systems in conjunction with their program executive officers and program managers.

Service chiefs have become increasingly disconnected from the acquisition process.

A large impetus for the schism was the Goldwater-Nichols Act passed in 1986, which was intended to improve unity of command and push the military departments and the Joint Chiefs of Staff to instill a more "joint" approach to military operations. Among its key provisions, the act established that only one office in a military department was to have responsibility for that department's acquisition process, and that office would report to the secretary of the military department (through a service acquisition executive). In effect, the Goldwater-Nichols Act put the service chiefs on the sidelines of the acquisition process.

Organizational and management issues contribute to inefficiency and unrealistic expectations in the acquisition process.

At the time, there was some concern about the impact of this step. For example, Senator Sam Nunn said he "was concerned that we not create an impenetrable wall between the staffs of the service secretary and the service chief" (1985). As time has passed, it does appear that the service secretaries and the service chiefs, along with their respective staffs, are not regularly engaging in meaningful dialogue.[1] Without such dialogue,

---

[1] For example, one program executive officer told the authors that "they have substantive discussions of my programs at the Pentagon, but I am not invited."

service chiefs may be intractable in emphasizing warfighting needs at the expense of reducing cost where possible; on the other hand, the acquisition process loses the operational insight that is critical in analyzing trade-offs between cost, schedule, and performance.

To mitigate this problem, the military department acquisition instructions for each military department could be modified to *explicitly* state the duties and expectations of the service chiefs and their staffs. As an example, these instructions could state that program decision meetings, where acquisition decisions are made, could be chaired by the service acquisition executive and the service vice chief. Such a revision would send a broad message to those who develop warfighting requirements that a link exists between the requirements process and the acquisition process.

The Combatant Commands' Role in Defense Management

There is a prevailing theory in Washington, D.C., that acquisition outcomes would improve if the combatant commands had a greater role in DoD's process for *determining warfighting requirements*. According to this theory, acquisition outcomes would improve because the military departments would better attune themselves to the warfighting capabilities that COCOMs need to have as well as when they need to have them; on the other hand, acquisition outcomes would also improve because the COCOMs would have to take fiscal constraints into greater consideration when identifying the capabilities they want.

There is an historic precedent for this concept. The idea of addressing warfighter needs more directly through participation of the COCOMs in DoD's decisionmaking processes emerged in the middle 1980s during the Reagan Administration. At that time, Integrated Priority Lists (IPLs) were introduced to allow unified and specified commanders (at that time known as commanders in chief, or CINCs) to influence the direction of resource allocation and weapon systems acquisition by identifying their highest priority requirements and making recommendations for programming funds. In concept, IPLs were a compelling idea: the people who are in charge of operationally leading our nation's military would appear to be the logical ones to determine what they need on the battlefield.

In practice, however, this approach did not improve outcomes. The then CINCs did put together their lists, but each was quite different from the next, both in content as well as depth. Some were thought-out and actionable (e.g., "I need 40 more Advanced Medium-Range Air-to-Air Missiles in my Area of Responsibility"). Others, however, were vague, and it was unclear what action was needed to address the warfighting need. Even when various administrations invited these senior military leaders to Washington, D.C., to present their ideas, it was clear that they were neither attuned to the requirements process nor to the acquisition process and had not given sufficient thought to the content of their IPLs.

As the type of information in historical IPLs indicates, the combatant commands are warriors equipped to fight today's wars and to prepare for tomorrow's threats; they are not equipped for a major role in developing comprehensive, detailed, and thus directly actionable requirements for the DoD acquisition processes. This is quite understandable. The COCOMs and their staffs need to spend their tour of duty understanding their Area of Responsibility and their potential adversaries through training, exercises, and building country-to-country military relationships. Taking a major role in DoD management processes would diminish their ability to carry out that mission, and could lead to an excessive emphasis in acquisition on near-term warfighting needs at the expense of long-term capability planning as well as to multiple and potentially duplicate staffs, all of which would ultimately affect the acquisition process.

> **The current system of checks and balances between the COCOMs and the departmental headquarters are in an approximate balance.**

The COCOMs' input into the requirements and acquisition processes is important and necessary, but current Title 10 allocations of responsibility already reflect the need to divide this labor by establishing the military department secretaries and their service chiefs to support the COCOMs by assessing and addressing their requests in light of the funding available, the other demands of their individual departments (e.g., manpower and support), the ability of the public and private sectors to deliver what is possible, the condition of the industrial base, and the amount of risk in a program.

In short, the current systems of the operational commanders and the departmental headquarters already reflect that the COCOMs cannot do everything and that the United States does not want to distract the COCOMs from winning wars and achieving other military objectives and actions within their Area of Responsibility. It is the job of those in Washington, D.C., to reach out to the COCOMs and demonstrate that their needs are being addressed, rather than turn the process over to them.

Joint Duty Requirements Erode Operational Insights Within Acquisition Program Management

Efforts to instill greater "jointness" in military operations have had an unintended impact on the acquisition process, both diverting line officers away from management roles in the acquisition process and eroding the management credibility of those officers who do choose acquisition duty assignments. Prior to the Goldwater-Nichols Act (which required a military officer to serve in a joint duty assignment in order to achieve flag or general officer rank), officers could move between "line" operational duty assignments and acquisition duty assignments. After the Goldwater-Nichols Act, they could no longer do so as readily if they wanted to achieve flag or general officer rank in an operational role because of the limitations of time in a career. Consequently, they lacked the opportunity to develop a deeper understanding of the acquisition process by serving in acquisition duty assignments. Furthermore, those who chose to devote their energies to the acquisition realm faced erosion of their operational credentials and lacked credibility when, for example, it came to determining whether a particular performance requirement was truly needed.

Finally, as the number of officers serving in acquisition roles decreased, a sense emerged of the acquisition process "belonging" to the largely civilian material establishment, not the line officers. Interviews with program executive officers, senior requirements officials, and political appointees in the acquisition process have provided strong anecdotal evidence that the channels of communications between the services' requirements and resourcing organizations (managed by line officers) and the service acquisition organizations are breaking down, and each realm is making decisions on common programs without consulting the other.

To improve this situation, the new administration could seek legislative change to give key service acquisition positions the equivalence of joint duty when it comes to potential promotion to flag or general officer rank. It would then become much easier for officers to move fluidly between operational and acquisition duty assignments, increasing the linkage between these two realms and also increasing the range of knowledge and expertise both of officers participating in the acquisition process and those serving to identify military requirements. Given the pressure on DoD's acquisition system, accomplishing such a change does not appear to be impossible.

Too Much Emphasis on Management Processes over Creativity and Initiative

Prescribing every step in the acquisition process has led to complexity, rigidity, and delay—not cost savings. The instruction that governs the department's acquisition process (DoD 5000.2) has been under constant review and modification from the late 1980s to today, with major updates issued in 1987, 1991, 1996, 2000, 2003, and 2008. The 2008 update references 79 laws, instructions, and directives, some addressing such esoteric matters as "Transfer Syntax for High Capacity ADC Media" and "Management of Signature Support within the Department of Defense."[2]

---

[2] For example, the Defense Procurement and Acquisition Policy (DPAP) office stated that "The UID PMO would like to announce the October 4, 2006, publication of ISO/

Recent acquisition reforms, which may have individual merit, together add up to a series of process changes that have reduced management flexibility and created new layers of bureaucracy. Some of these reforms, implemented under the guidance of former Under Secretary of Defense for Acquisition, Technology, and Logistics (USD (AT&L)) Young with the best of intentions, and at times codified into law by the Congress, include

- competitive prototyping[3]
- a new series of Joint Action Teams
- Capability Portfolio Management Teams
- Reliability, Availability, and Maintainability Policy
- defining the "desirable attributes" of the defense industrial base and the methodology to assess industry progress toward developing these attributes
- Configuration Steering Boards (now in law)
- Joint Rapid Acquisition Committee.

The Joint Staff has also introduced changes that have increased the complexity of the overall acquisition process, replacing the Requirements Generation System with the Joint Capabilities Integration and Development System (JCIDS) in 2003. With JCIDS came a new series of analytic tools and documents. Many of these reforms involve the rote application of rules and the production of prescribed management information in preset templates that may not necessarily be appropriate or applicable for a specific program. Figure 4.1, which

Prescribing every step in the acquisition process has led to complexity, rigidity, and delay—not cost savings.

IEC 15434.3, Transfer Syntax for High Capacity ADC Media, which establishes '12' as the approved Format Code for Text Element Identifiers." As of May 14, 2009: http://www.acq.osd.mil/dpap/UID/enewsletter/archive/dec06/.

[3]   See, for example, Drezner, Jeffrey A., and Meilinda Huang, *On Prototyping: Lessons from RAND Research*, OP-267-OSD, for an in-depth discussion of the benefits and issues of competitive and other prototyping.

Figure 4.1
DoD's Requirements, Acquisition, and Funding Processes

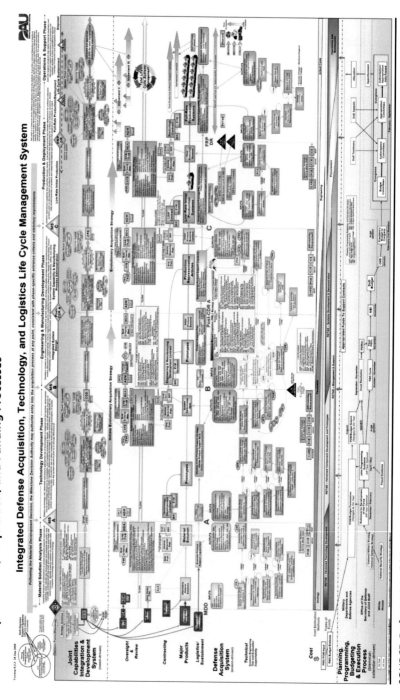

SOURCE: Defense Acquisition University, 2008.

contains DoD's representation of the acquisition management framework, speaks for itself in terms of the complexity of the department's requirements, acquisition, and budgeting processes.

From the perspective of the acquisition professionals in the Office of the Secretary of Defense (OSD), in the military department secretariats, and in the program manager and program executive offices across DoD, the extent to which these changes govern their actions leads to only one conclusion: Success means getting a program through the wickets—not the application of logic, cost-cutting measures, or technology. Of course, their job *should be* to think, create, or seek efficiencies, but the process motivates them instead to follow rules in a rote manner. Changes the Clinton administration made to contracting rules in the middle 1990s shed some light on the impact of DoD's emphasis on management processes. That administration eliminated all specifications and standards, but the acquisition workforce (AW) and the defense contractors did not embrace this new freedom, preferring the safety of highly prescribed procedures they had become accustomed to. So, while military specifications were not required, defense contractors translated them into their own "company" specifications and used them on contracts, knowing that the government would readily accede to them.

One reason for DoD's growing emphasis on management processes is its decreasing supply of knowledgeable acquisition professionals. This decrease has occurred, in part, due to the enactment of false economies. For example, a critical event that helped shape the defense acquisition environment—the Defense Management Review process—had unforeseen long-term consequences. The Defense Management Review attempted to derive economies in defense spending by enhancing the efficiency of all elements of DoD through better material management practices, better organizational efficiency, and elimination of staffing redundancies. The review anticipated billions of dollars in savings from these efforts, and eliminated funding and billets from DoD's program plan. Many of the positions eliminated were in the acquisition management and oversight areas. Instead of simplifying the acquisition processes commensurate with a reduction in staff, the effort tried to substitute management processes for a cadre of knowledgeable engineers, logisticians, and material managers, decreasing creativity in our acquisition process and forcing DoD to rely on additional support contractors to execute the process. We believe that doing so has created burdens that are neither well understood nor sufficiently evaluated. Clearly, the acquisition process is tending toward a bookkeeping process where innovation is not accepted or encouraged. The nation needs to recapture, motivate, and reward the competitive drive and the innovation of its acquisition professionals.

A more streamlined requirements and acquisition process needs to be put into effect, but it should be one that encourages workforce initiative and responsibility. Federally funded research and development centers (FFRDCs) and others could play a role in the solution, but there is a need to find the best practitioners, both uniformed and civilian, and empower them to identify the necessary changes. Laws and regulations will need to be reassessed to determine the latitude available, but with Congress complaining about the cost outcomes of the current system, even legislative change should be achievable. There are enormous potential savings associated with this change, but unshackling key military officers and civilian personnel by allowing them to bring their creative powers to bear on spiraling acquisition costs is essential to give affordability more priority in defense acquisitions. To start, the elimination of duplicate processes in the military departments and the OSD would be a positive step, with the former focusing on execution and the latter on establishing policy. Better lines of decisionmaking authority would be a positive step as well, along with truly streamlining the acquisition; requirements; and planning, programming, budgeting, and execution (PPBE) processes.

Conclusions

In this paper, we have highlighted several organizational and management issues, the resolution of which could have a key role to play in improving acquisition outcomes. All of the problems facing defense acquisitions cannot be placed on their doorstep, but attending to these issues properly will lead to a more efficient, integrated, and innovative basis for decisionmaking, without which improved acquisition outcomes will not be possible. To this end, we offer the following recommendations:

- **Increase the service chiefs' role in the acquisition process.** In order for the requirements, funding, and acquisition processes to better function together, the service chiefs need a central, but not controlling, voice in the acquisition process. Having the service vice chiefs as co-chairs of the Military Department's Acquisition Board would be a step in the right direction. It would give the service chiefs a greater sense of participation in the acquisition process.
- **Ensure that the combatant commanders' role in the Pentagon's resource allocation processes is strictly advisory.** The input from the combatant commanders is very important, but they need to be free from bureaucratic processes that would take too much time away from their warfighting responsibilities.
- **Make acquisition positions equivalent to joint duty.** To entice the military officer line community to opt for acquisition positions, make a series of these positions the equivalent of joint positions, thus better enabling these officers to compete for higher rank.
- **Emphasize workforce initiative, not management processes.** Methods need to be developed—and training and incentives need to be constructed—to enable the acquisition workforce to think for itself, as opposed to the current process of prescribing everything through an instruction or regulation.

CHAPTER FIVE

On Prototyping: Lessons from RAND Research

Jeffrey A. Drezner and Meilinda Huang

Introduction

Acquisition policy and practice reflect the recurring theme that prototyping as part of weapon system development can improve program outcomes. Specifically, prototyping is widely believed to reduce cost and time; allow demonstration of novel system concepts; provide a basis for competition; validate cost estimates, design, and manufacturing processes; and reduce or mitigate technical risk.[1] A mandate for competitive prototyping has periodically been included in revisions to the DoD 5000 series of regulations governing the defense acquisition system. The most recent revision to DoD Instruction 5000.02 (December 2008, p. 17) contains the following mandate:

> The TDS [Technology Development Strategy] and associated funding shall provide for two or more competing teams producing prototypes of the system and/or key system elements prior to, or through, Milestone B. Prototype systems or appropriate component-level prototyping shall be employed to reduce technical risk, validate designs and cost estimates, evaluate manufacturing processes, and refine requirements.

This paper reviews four decades of RAND research on the uses of prototyping in DoD in order to draw lessons that practitioners can

---

[1] See, for instance, Young, 2007.

apply as they respond to the new emphasis on prototyping in DoD acquisition policy.[2] RAND's research on this topic has been periodic, reflecting the waxing and waning of DoD interest in prototyping. We make only limited use of prototyping-related studies outside of RAND research; these studies have also been episodic in their coverage of prototyping.

In this paper, we first define prototyping and discuss the potential benefits commonly attributed to it. We then review historical analyses to determine how well those potential benefits are supported by experience. Last, we make explicit some of the more important thematic lessons and specify the conditions under which prototyping is most likely to yield expected benefits.

What Is Prototyping?

In general, prototyping is a set of design and development activities intended to reduce technical uncertainty and to generate information to improve the quality of subsequent decisionmaking. Although the term "prototyping" captures a wide range of activities, all prototyping has several elements in common, including the design and fabrication of one or more representative systems (hardware or software) for limited testing and demonstration prior to a production decision. The prototype itself is the article being tested.

> **Prototyping is a conscious strategy to obtain certain kinds of information to inform specific decisions.**

A prototype is a distinct product (hardware or software) that allows hands-on testing in a realistic environment. In scope and

---

[2]   Of approximately 30 reports and papers reviewed, eight were specifically focused on prototyping; the others touched on aspects of prototyping in the course of exploring other acquisition issues, such as development strategies more generally or cost and schedule issues.

scale, it represents a concept, subsystem, or production article with potential utility. It is built to improve the quality of decisions, not merely to demonstrate satisfaction of contract specifications. It is fabricated in the expectation of change, and is oriented towards providing information affecting risk management decisions. (Drezner, 1992, p.9)

In other words, prototypes are distinct from full-scale, final configuration, production-representative test articles, the main purpose of which is to verify that performance requirements have been met and the program is ready to move into the production phase. The expectation that prototyping will lead to change implies that a prototype is intended as a vehicle to learn something about the system's technology or concept that will inform subsequent decisions. The prototype itself does not have to be a fully configured production article to accomplish this purpose.

Prototyping can take many forms. It can be conducted at both the system and subsystem level. It can include competition (e.g., two or more teams designing, fabricating, and testing a representative system in the context of a source-selection decision) or just a single organization experimenting with a novel concept or new technology. Some developmental activities (i.e., experimentation, system concept and technology development, demonstration and validation) are often not labeled as prototyping, but the nature of the activities planned and accomplished is consistent with prototyping. For example, Advanced Technology Demonstration (ATD), Advanced Concept Technology Demonstration (ACTD), and Joint Concept Technology Demonstration (JCTD) projects involve specific kinds of prototyping activities.[3]

Prototypes can be part of the early stages of a Major Defense Acquisition Program (MDAP), part of a series of related efforts (e.g.,

---

[3] See Drezner, 1992 for a full exploration of the different kinds of prototyping strategies.

the X-plane series of experimental aircraft)[4] or stand-alone projects. However, it is important to note that prototyping alone does not constitute a full weapon system development program. When incorporated into an MDAP, prototyping should be used together with design analysis, empirical testing, modeling and simulation, and "other methods of reducing technological uncertainty" (Perry, 1972, p.41) because these other methods produce information that prototyping alone will not; such approaches are complementary in the context of an MDAP. When prototyping is done outside of an MDAP, the transition and transfer of technology and information becomes an important practical issue affecting the value of the prototyping effort. Unlike an MDAP, which has a constituency who ensure the political and budgetary support necessary to move the project forward, an ATD or ACTD program does not necessarily have the support of the military service for which it ultimately may be intended.[5]

Some weapon types are generally too costly to prototype at the system level (e.g., large naval surface combatants and complex satellites). In such cases, *subsystem* prototyping is a cost-effective alternative for reducing uncertainty. For instance, the DD(X) program (now called DDG 1000, the Navy's newest guided missile destroyer) successfully used a series of engineering development models (EDMs)—that is, prototypes of critical subsystems, such as the hull form, advanced gun and its munitions, composite deck house, peripheral vertical launch missile system, and radars, among others—to reduce technical risk and refine subsystem design.[6]

---

[4]   The X-planes were experimental aircraft (from the X-1 in 1946 to the X-53 in 2002) designed to expand knowledge of aerodynamics and air vehicle and engine design. Individual projects were run by combinations of NASA, the U.S. Air Force and U.S. Navy, and industry. The X-1 was the first aircraft to break the sound barrier. The X-15 achieved hypersonic (Mach 6) manned flight. Other projects demonstrated different wing or body configurations. The first unmanned combat air vehicles (X-45 and X-47) were industry projects intended to demonstrate a new capability.

[5]   The Predator and Global Hawk unmanned aerial systems (UAS) programs are good examples of this issue and also demonstrate that the issue can be resolved. See Thirtle, et al., 1997; Drezner and Leonard, 2002.

[6]   Several U.S. Government Accountability (GAO) reports discuss the role of the EDMs in the overall program, including GAO-07-115 (November 2006), GAO-05-924T (July 2005)

Although the Under Secretary of Defense for Acquisition, Technology, and Logistics (USD (AT&L)) defined prototyping activities as occurring prior to Milestone B (Young, 2007, p.1), and the recently revised DoD acquisition policy, quoted above, places prototyping in the technology development phase (before Milestone B and the start of engineering and manufacturing development), prototyping as defined in this occasional paper is not confined to any particular acquisition-process stage of the development process. Moreover, our definition covers a broader range of activities and is not confined to a particular funding mechanism or program type.

What Are the Expected Benefits of Prototyping?

Theoretical arguments heavily favor the use of prototyping as part of weapon system development, or more specifically, as a way to demonstrate novel concepts and resolve technological uncertainty. The expected benefits, as listed below, are largely related to the design, fabrication, and test of the prototype. However, prototyping can be applied in any situation in which improved information through demonstration would be of value. Less traditional applications include prototyping specific management techniques or policies, or prototyping the support and maintenance concepts for a weapon system.

Given DoD's recent competitive prototyping policy, it is useful to compare DoD's list of expected benefits with similar sets of benefits and skepticisms generated in prior research by RAND and others. DoD policy lists the following primary benefits:

- Reduce technical risk.
- Validate designs.
- Validate cost estimates.

---

and GAO-04-973 (September 2004). Full citations are in the Reference list at the end of this paper.

- Evaluate manufacturing processes.
- Refine requirements.

The policy directive on competitive prototyping (Young, 2007, p. 1) listed these as well as a set of secondary benefits:

- Exercise and develop government and industry management teams.
- Provide an opportunity to develop and enhance system engineering skills in both government and industry.
- Attract a new generation of scientists and engineers to DoD and the defense industry.
- Inspire and encourage students to embark on technical education and career paths.

The expected benefits of a particular prototyping strategy for an individual program will not necessarily include all of these potential benefits. Front-end planning activities should identify which benefits are being targeted and then design a prototyping strategy to provide those specific benefits. In other words, prototyping involves a conscious strategy to obtain certain kinds of information to inform specific decisions. In the remainder of this section, we examine expected benefits of prototyping in more detail.

Reduce Technical Risk. Reducing technical risk is an important benefit of prototyping strategies. By building and testing representative items, prototyping can identify and resolve technical uncertainty; demonstrate technological feasibility; advance technological maturity; refine system requirements and validate the system design to satisfy those requirements; and provide information to improve estimates of cost, schedule, and performance.

> **Prototyping can resolve known technical uncertainties— and identify uncertainties that were not anticipated ("unknown unknowns").**

An important aspect of risk reduction is the discovery of technical uncertainty not anticipated by the design engineers nor predicted by design analyses or prior experience with analogous technologies. In other words, prototyping can address both the "known unknowns" and the "unknown unknowns," sometimes yielding unexpected (unlooked-for) benefits. For instance, testing of the Global Hawk (an unmanned aerial vehicle) resulted in the development of an unplanned for capability (in-flight retasking)[7] not envisioned at the design stage (Drezner and Leonard, 2002).

Validate Design and Refine Requirements. Prototyping strategies are fundamentally about using information generated during design, fabrication, and testing activities to inform subsequent decisions regarding cost–performance trade-offs, source selection, validation of technologies, readiness to move into subsequent program phases, and force structure. Relatively more and higher-quality information can be generated during prototyping than in alternative development approaches that rely more heavily on "paper" design activities or, more recently, on computer-aided design, modeling, and simulation. Presuming that the information that is generated is used appropriately, the quality of those decisions should improve.

Validate Cost Estimates. While prototyping may not truly validate cost estimates (most prototyping occurs early in development, and the prototypes themselves are not the full-production configuration), it may improve the quality of those estimates. Other benefits discussed—such as reductions in technical risk, more mature technology, refinement of requirements based on demonstrated feasibility, and validation of key system design elements—should enable a more accurate cost estimate.

Evaluate Manufacturing Processes. A prototyping strategy can be designed to evaluate or improve manufacturing processes. Achieving this potential benefit would require making elements of the manufacturing process (i.e., tooling, material use and handling, production process layout) an explicit part of the prototyping strategy. There is

---

[7]   That is, redirecting the air vehicle so onboard sensors can capture targets of opportunity.

an important trade-off here: Including the full set of production considerations in a prototype designed, fabricated, and tested as part of technology development would likely add cost and time to the effort. Nevertheless, the construction of the prototype itself will provide some valuable insights to evaluate and improve production processes.

The expected secondary benefits identified by DoD policy (listed above) relate either to the development and maintenance of program management, system engineering and other technical skills, or to the development and recruitment of the next generation of defense program managers, scientists, and engineers. RAND research has addressed the former but has not addressed the latter.

Maintaining Workforce Skills and Experience. Prototyping helps to sustain industry design capabilities through design, fabrication, test, and redesign activities. Prototyping provides a more complete experience for a design team. As past RAND reports have noted:

> To be really good at designing combat aircraft, members of a design team must have had the experience of designing several such aircraft that actually entered the flight-test stage. Paper designs and laboratory development are important, but they are not a substitute for putting aircraft through an actual flight-test program. (Drezner, 1992, p. 16)

> If experience is as important as might be inferred from the historical record, clearly the DoD needs to consider options that will help maintain experience levels during long periods when no major R&D programs are under way. Such a strategy could focus on prototyping or technology demonstration. (Lorell, 1995, p. 65)

Similarly, prototyping activities may provide the government workforce with hands-on experience in program management, system engineering, testing, and other skill sets necessary for the conduct of a successful acquisition program.

RAND research suggests additional potential benefits from prototyping not explicitly listed in DoD policy, including the following.

Reduction in Fielding Time. Prototypes can demonstrate the military utility of a system concept or technology, enabling relatively shorter time spans from concept definition to fielding of an operationally useful capability. This applies to prototyping both as part of a MDAP (e.g., the YF-16 Lightweight Fighter, which led to the F-16A/B) or stand-alone, pre-MDAP programs (e.g., fielding the ACTD configuration of Global Hawk during the initial stages of the wars in Afghanistan and Iraq).

Enhanced Competition. Prototyping strategies can enhance competition among two or more firms or teams by requiring actual demonstration of proposed capabilities. Competitive prototyping has been used extensively, for example, in the history of fighter and bomber aircraft development, often demonstrating the value of new designs or technologies and giving government source-selection authorities increased confidence in their decisions (Lorell, 1995). The testing of competitive prototypes can provide better information than design proposals alone.

Improved Research and Development (R&D) Efficiency. A prototyping strategy can also be more efficient, providing an opportunity for "obtaining information sooner or more cheaply than by other means" (Perry, 1972, p. 41).

Hedging Your Bet. Some prototyping strategies can offer a hedge against other kinds of uncertainty beyond technical uncertainty. For instance, a competition with two or more system designs provides a hedge against the nontechnical failure of one. For example, technology demonstration prototyping strategies—in which system concepts are tested outside of established programs—can provide a hedge against changing or emerging threats. Within DoD, these programs are usually ATDs, ACTD, JCTDs, or similar programs.

Skepticism about the benefits of prototyping is less common than enthusiasm, but it does exist. The counterarguments revolve around two notions—that changes in performance requirements (capabilities) and duplication of effort reduces the value of prototyping. The first notion is that a prototype phase does not really reduce uncertainty (or risk) because decisionmakers will be unable to resist the temptation to modify system performance specifications to capitalize on recent technological advances. Incorporating that new technology will increase risk since those changes were not part of the prototype phase, thus reducing the value of prototyping. The second notion is that:

> a comprehensive design effort is unavoidable in any case and . . .
> pausing . . . to construct a prototype merely lengthens the program
> and increases its cost without securing any equivalent benefits.
> The argument is that engineering problems will be encountered in
> both cases, but that careful study and design analysis will identify
> them earlier than will prototype construction. Furthermore, it is
> widely believed that the construction of a prototype encourages
> designers to overlook compatibility problems, to create something
> that is less than a system and that must be substantially reengi-
> neered before it is ready for production. (Perry, 1972, p. 10)

The counterargument about changing requirements after prototype testing revolves around the notion that significantly changing the design of a system reduces the value of the information obtained through prototyping. This may be a valid concern at the extreme, where design changes subsequent to prototype testing result in a completely different configuration. However, one could also argue that the prototyping experience resolves uncertainties associated with the initial requirements, a nontrivial contribution to the development program even if requirements are changed somewhat or new capabilities added. To an important degree, prototyping is intended to result in design changes based on lessons from testing. It may also identify flawed or infeasible requirements, to the benefit of the program. Prototyping does not resolve all uncertainties associated with a system or technology concept, but rather only those it was designed to resolve and perhaps some "unknown unknowns" that become apparent during testing.

> **Skeptics contend that subsequent changes to performance requirements and duplication of effort reduce the value of prototyping.**

The second counterargument—that the prototyping effort is duplicative and produces little unique knowledge informing the detailed design phase—is less valid. For instance, even with advances in computational fluid dynamics, wind tunnel testing and live flight testing of aircraft configurations are still required for designs that push the edges of known and demonstrated performance, as many military systems do.[8] Demonstrations in realistic operational environments consistently produce information about system performance not otherwise obtainable, and fabrication of a prototype certainly exercises skills that design activities alone cannot. Prototype testing also enables identification of any unknown or unexpected performance behaviors or technical risk.

Historical Evidence Is Mixed

RAND's past research on the topic of prototyping includes both (1) statistical analyses of large databases containing information on both prototyping and nonprototyping programs and (2) case studies (in varying degrees of detail) of prototyping programs, with comparisons to nonprototyping programs. The programs studied in this body of research include the following, among others:

- Century-series fighters (F-100 through F-105)
- AX close air support/attack aircraft (YA-9 versus YA-10)
- Lightweight Fighter (YF-16 versus YF-17)
- Advanced attack helicopter (YAH-63 versus YAH-64)
- Utility transport helicopter (UH-60 versus UH-61)
- F-117 (Have Blue)
- Advanced Medium-Range Air-to-Air Missile (AMRAAM)
- Predator unmanned aerial system
- Global Hawk unmanned aerial system

---

[8] For a discussion, see Antón, et al., 2004, both TR-134-NASA/OSD and MG-178-NASA/OSD.

- DarkStar unmanned aerial system
- F/A-18 and F/A-22 fighter aircraft

Overall, evidence from this body of work is somewhat mixed regarding the benefits of prototyping. However, many factors affect program outcomes independent of prototyping; thus, teasing out the relative effect of prototyping is challenging.

In general, we would expect to find that programs incorporating prototyping would have less cost growth on average because the baseline cost estimate would benefit from the risk reduction and information on cost–performance trade-offs obtained through early prototyping.[9] Findings from numerous case studies support this expectation, as indicated in the following examples:

> The experience of the Air Force in buying "soft tooling" prototypes, including the two XF-104s, suggests that under appropriate conditions an airframe very useful for flight testing of both basic designs and readily available subsystems might be obtained for about 60 percent of the cost of a "hard tooled" prototype. And of course it becomes available much sooner. (Perry, 1972, p. 39)[10]

> There is some evidence that, on average, cost growth of prototyped programs is less than that of conventional acquisition programs, and the magnitude of such "savings" is much greater than the direct cost of the prototype phase. (Smith et al., 1981, p. 35)

---

[9]   In this paper, cost growth was measured from the Development Estimate baseline established at Milestone II (now called Milestone B) using quantity adjusted data from the Selected Acquisition Reports (SARs). Though this analysis is 15 years old, RAND has continued to update the database. More recent analyses, while not explicitly addressing prototyping, are consistent with the basic findings in the earlier report and do not provide any indication that results on prototyping might change. See also Arena et al., 2006; Younossi et al., 2007; and Bolten et al., 2008.

[10]   "Soft tooling" consists of the temporary set of tooling used to construct a limited number of prototypes. Such tools (molds for shaping materials, presses and drills, wire fitting, workstations, et cetera) may adapt general purpose or existing tools, or may even be made of wood. "Hard tooling" refers to the final set of tools used for longer production runs.

Figure 5.1
Cost-Growth Factor Distribution of Prototyping and Nonprototyping Programs, Circa 1993

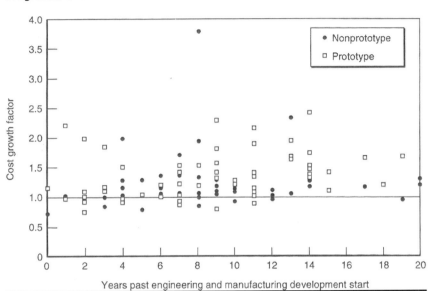

SOURCE: Drezner et al., 1993, p. 37, Figure 5.1.

However, a statistical analysis conducted in the early 1990s of the factors affecting weapon system cost growth—including the effect of prototyping on cost outcomes—found no easily discernible patterns in the data (see Figure 5.1).[11] If anything, the data appeared to indicate that prototyping programs had higher cost growth than nonprototyping programs, with average cost-growth factors of 1.29 and 1.19,[12] respectively (there were 30 observations in both groups of programs).

This counterintuitive result can be partially explained by the fact that the programs in the prototyping group were, on average,

---

[11]   Drezner et al., 1993.

[12]   Cost-growth factors translate directly to percentages: an average cost-growth factor of 1.29 indicates that on average, the programs in that group had 29 percent cost growth above their baseline estimates.

older and smaller than the programs in the nonprototyping group: Relatively older (more mature) programs (measured as years past the baseline cost estimate) tend to have higher cost growth, and relatively smaller programs (measured in inflation adjusted dollars) also tend to have higher cost growth. However, when we compared programs in which the prototyping phase occurred earlier—prior to Milestone II (when the program baseline was established)—to programs in which the prototyping phase occurred after the milestone, the expected result obtains: programs with earlier prototyping had an average cost-growth factor of 1.23 versus 1.37. This result indicates that using the information generated during early prototyping may improve subsequent cost estimates.

> A statistical analysis suggested that programs that use prototyping do not show less cost growth— unless prototyping occurs prior to Milestone II.

Another widely held belief is that spending relatively more time prior to Milestone B (formerly, Milestone II) in planning and technology demonstration activities (including prototyping) would result in less schedule slip. A study that examined in detail the factors affecting schedule slip of 10 MDAPs did not find evidence to support this hypothesis (Drezner and Smith, 1990); however, several other studies did. For example:

There is no evidence that the introduction of a prototype either delays the availability of the production article or increases the cost of development. (Perry, 1972, p. 45)

The histories of attack and fighter aircraft developed by the Navy and the Air Force since 1950 indicate that introducing a prototype makes little difference in the total development time. Furthermore, if a prototype program can be started earlier than could an equivalent full-scale development program (as was certainly the case with the LWF program), then use of a prototype phase may actually lead to an earlier fielding date. (Smith et al., 1981, p. 35)

In addition, one study that compared the development strategies and outcomes of two programs using differing prototyping strategies found evidence that prototyping benefited both programs:

> Another major difference between the two programs was that the YF-16 was a true prototype of the F-16 flight vehicle, thereby providing a considerable start on the overall system design. Conversely, the Have Blue program was a technology demonstrator and provided almost nothing toward the detail design of the F-117. On that basis, we would expect that, measured from EMD start, the F-117 schedule would have been extended, compared with the F-16, while in fact the time to first delivery was about the same for both programs. This suggests that the F-117 program was relatively short. (Smith et al., 1996, p. 30)

Though it was a subscale technology demonstrator, Have Blue did validate key aspects of the F-117 design, providing increased confidence to decisionmakers. The two programs had very different prototyping strategies, but both appear to have derived important benefits, including the generation of information that facilitated a relatively shorter development phase. Prototyping activities provide benefits by generating information not otherwise attainable.

One possible explanation for these mixed results concerns the exact metric used in the analysis. By their nature, prototyping strategies generate useful information applicable mainly to a particular design, technology, system concept, or other engineering challenge. The information would be expected to improve decisionmaking or estimates for systems closest to the prototype's design. For instance, when measuring the effect of prototyping on the accuracy of cost, schedule, or performance estimates, initial estimates of cost, schedule, or performance should be compared only to the version of the system that is based on the prototype. Applying this principle to F-16 program outcomes, discussed in the last case study above, only the F-16A/B models should be considered in the analysis; the F-16C/D models came much later, and cost, schedule, or performance estimates were not likely affected by the prototyping experience of earlier years.

Unfortunately, this level of discrimination in a study is difficult to achieve due to the limits of available data and so has been undertaken only infrequently.

Perhaps more importantly, many factors other than prototyping affect program outcomes. Such factors include infeasible requirements, requirements change, budget instability, and underestimation of technological maturity. As discussed above, prototyping can be designed to address the feasibility of requirements and technological maturity, but any benefits in these areas can be overwhelmed by other factors. The F-22, the JSF (F-35), and Global Hawk all included prototyping strategies of one form or another that appeared quite useful at the time, yet all three programs have incurred high cost growth and schedule slip.[13] This suggests that prototyping is not a panacea for solving all of the problems of the acquisition process.

In general, evidence from case studies tends to support the notion that prototyping strategies are beneficial as part of weapon system development in some circumstances. Prototyping does help discover technical risks and thus can reduce technical uncertainty. Prototyping does produce information useful in validating design choices, refining requirements, and improving the quality of cost estimates. However, results from both case studies and statistical analyses suggest that the impact of these benefits on cost, schedule, and performance outcomes can be overwhelmed by other factors affecting programs.

Conditions That Favor Prototyping

Though the available evidence is somewhat mixed overall, the historical record does suggest some of the conditions under which prototyping strategies are most likely to yield benefits in a development program. Successful application of prototyping strategies in the future requires either creating these conditions or ensuring that they exist to the extent possible.

---

[13] See Selected Acquisition Reports for the programs. See also several GAO reports.

Results Are Used to Inform Key Program Decisions. If the information generated from prototyping activities is not used to inform key program decisions (including final design, technologies and capabilities to include in the initial production system, planning for subsequent technical and engineering activities, and cost and schedule estimates), then there would be no reason to expect benefits. If early testing of a prototype indicates that available technology is not yet mature enough to confidently predict that system performance requirements will be met, then pushing ahead in that program without easing requirements and performance expectations to match demonstrated technological maturity will result in significant cost growth, schedule slip, and performance shortfalls.

The Prototype Is Designed to Demonstrate the Critical Attributes of the Final Product in a Realistic Environment. Prototypes should be designed to test the key performance attributes about which there is the greatest uncertainty and which are expected to enable mission accomplishment. This includes major subsystems that affect not only performance, but also design (such as the integration of a specific engine, airframe configuration, and sensor package in an aircraft). Prototyping strategies appear to yield benefits when they are focused on specific challenges or designed to generate specific kinds of information to inform specific kinds of decisions

> **Prototyping is more effective when critical performance attributes are tested in a realistic environment.**

Prototyping Strategies and Documentation Are Austere. There is some evidence, particularly from the many past aircraft prototypes, that an austere program is an important attribute of a successful application of prototyping. Prototyping should include only the minimum necessary requirements specified and only the minimum documentation required to analyze test results and capture lessons learned from the activity. In general, this means focusing the prototype itself on a few key uncertainties, keeping noncritical technical standards to a minimum, and focusing associated program documentation on the prototyping activity. It also means the use of relatively small teams of highly capable people with appropriate decision authority regarding the prototyping activity, minimal requirements for status reporting, and minimal external interference (e.g., externally imposed design changes) with the team's activities. This gives the design team more flexibility to make the inevitable cost–performance trade-offs, such as deciding not to include demonstration of a second-order capability due to cost considerations.

Sustainment issues, technical data requirements, production planning, and tool design are commonly not addressed in an austere prototyping strategy. However, these issues could be addressed through a phased or incremental prototyping strategy in which two sets of prototypes are designed and tested—the first addressing critical technical performance issues and the second addressing support issues. While possible in theory, we are not aware of any program that has attempted such an approach.

There Should Be No Commitment to Production During the Prototyping Phase. Prototyping is experimental in nature, and failure is a possible outcome in the sense that the desired capabilities could not adequately be demonstrated in a realistic environment and at a reasonable cost. Such an outcome would be strong evidence that the requirements need to be relaxed and additional technology development and maturation is needed. Perhaps the program based on the system concept should be deferred indefinitely until certain critical technologies are demonstrated. Such decisions are much harder to make if a commitment to production has already been made, either implicitly or explicitly. Production requires that a whole other set of issues be addressed (force structure, sustainment options, significantly increased budgets, et cetera).

> **Prototyping may lead to tough decisions that are best made before committing to production.**

No Additional Requirements Are Added or Performance Increases Expected. Changing the design to add capabilities that were not part of the initial design concept and therefore not explored during prototyping may limit the value of the information gained during prototyping. Again, this condition relates to the need for an austere, focused prototyping effort in which the information generated is used to inform specific decisions regarding design, requirements, and technology.

Conclusions

A careful application of prototyping can result in significant benefits to a program, including reduction in technical risk and demonstration of technological feasibility, refinement of requirements, and more informed cost–performance trade-off decisions. However, prototyping alone cannot ensure a successful program outcome; cost, schedule, and performance outcomes are affected by a range of factors independent of prototyping that may overwhelm any benefits gained through prototyping.

DoD's new acquisition policy mandates competitive prototyping at either the system or subsystem level prior to Milestone B.[14] Competitive prototyping is one specific kind of prototyping strategy involving two or more teams designing, constructing, and testing their respective system (or subsystem) and technology concepts. This type of prototyping strategy usually happens relatively early in the technology development phase of a program. The prototypes themselves are usually limited to demonstrating specific design concepts and technologies, and can provide information not otherwise attainable to inform the source-selection decision. This meets the definition of an austere prototyping strategy and satisfies the conditions discussed above as facilitating success. To the extent that the resulting information is properly used to inform program decisions at Milestone B, and no additional requirements or capabilities are added after the baseline is established, the policy may contribute to an improvement in program outcomes. However, recent experience represents a cautionary tale: The F-22, JSF,

---

[14]   See DoD Instruction 5000.02, December 2008.

and LCS all included a competitive prototyping phase, and all have experienced cost growth and schedule slip.

There are several important caveats regarding the potential of competitive prototyping that acquisition officials should consider. First, the competitive aspect of this policy requires two or more teams with the requisite knowledge and capabilities at the system or subsystem level. However, consolidation in many sectors of the industrial base has changed the nature and value of competition.[15] In these cases, competition will not necessarily yield the benefits of innovation and of cost reduction and control that are usually expected.

Second, by mandating competitive prototyping in the technology development phase, DoD's policy may inhibit other prototyping strategies. The successful application of any prototyping policy requires that officials think through the goals, acquisition environment, technical characteristics, and needs of a given program to determine what type of prototyping makes sense. The policy mandate might result in officials forcing a competitive prototyping strategy into the design of a program when the characteristics of that program require some other approach to addressing risk. For instance, the demonstration of the military utility of a new concept or technology does not always require competition. The discretion and judgment of experienced program managers and oversight officials are important conditions for successful implementation of this new policy mandate.

Last, the lack of definitive evidence supporting the benefits of prototyping in general, and competitive prototyping in particular, is somewhat troubling. Existing case studies and statistical analyses present the policymaker with mixed results. As a result, DoD's new competitive prototyping mandate was incorporated into policy without a strong link between the new policy emphasis and its intended improvement to program cost, schedule, and performance outcomes.

---

[15]   In particular, if there are only two firms (or teams) that can design and build a particular system or subsystem, and there is a formal or informal policy to maintain at least two, then competition is very different than it was in the past. Shipbuilding, manned aircraft and helicopters, and heavy armored vehicles are sectors where this concern is real. See Schank et al., 2006; Birkler et al., 2001; and Birkler et al., 2003.

Does competitive prototyping really result in better outcomes? Under what conditions will competitive prototyping yield the desired benefits? What are the key lessons from past and more recent experience with competitive prototyping? How can the potential benefits of competitive prototyping be maintained in the face of all the other factors affecting program outcomes? A carefully structured analysis of prototyping strategies, with an emphasis on recent experience with competitive prototyping (e.g., F-22, JSF, and LCS), would help ensure a more successful implementation of the new policy.

CHAPTER SIX

Shining a Spotlight on the Defense Acquisition Workforce—Again

Susan M. Gates

Introduction

As we approach the end of the first decade after the turn of the century, concerns about defense acquisition outcomes—cost escalation, reports of improper payments to contractors, appeals filed over source-selection outcomes, schedule delays—pervade the popular press as well as DoD audits and internal reports. Although the term "defense acquisition" refers to all activities that are related to the procurement of goods and services from the private sector by DoD, two specific types of acquisition activities are the source of greatest concern today: Major Defense Acquisition Programs (MDAPs) and contracting efforts to support immediate needs in a contingency or combat operation (often referred to as "expeditionary contracting"). The U.S. Government Accountability Office (GAO) has designated defense contract management and defense weapon system acquisition as "high risk" areas.[1] Another recent prominent assessment, the Report of the Acquisition Advisory Panel (Section 1423 Report) criticized government acquisition efforts for awarding a substantial number of contracts (nearly one-third) through noncompetitive approaches, and the Report of the Commission on Army Acquisition and Program Management in Expeditionary Operations (Gansler Commission Report) concluded that "the

---

[1] See the U.S. Government Accountability Office's Web site at http://www.gao.gov/doc search/featured/highrisk_march2008.pdf for a list of all GAO high-risk areas and the year in which they were designated as such; see also http://www.gao.gov/new.items/d07310.pdf on defense contract management, p. 71.

acquisition failures in expeditionary operations require a systemic fix of the Army acquisition system" (p. 1).

The cacophony of criticism is not new, echoing stories from the 1980s about the government spending inordinate amounts of money on everyday items such as toilet seats or hammers (Fairhall, 1987). Now, as then, critics have shined a spotlight on the acquisition workforce (AW)—its size, quality, and effectiveness—as a key contributing factor to the observed problems.[2] Indeed, a recent review conducted by DoD of "almost every acquisition improvement study . . . concluded in some fashion or another that more attention needs to be paid to acquisition workforce quantity and quality" (Lumb, 2008, p. 20). The following three workforce-related claims feature most prominently in the current debates:

1. The current workforce is too small to meet current workload. The Gansler Commission Report attributes poor contracting outcomes, including recent contracting scandals, to insufficient growth in the size of the contracting workforce and exploding growth in the acquisition workload (Gansler Commission Report, 2007, p. 30). This perspective is consistent with more general arguments that have been made about the federal AW overall, most recently in the Section 1423 Report, which stresses that the demands on the federal AW have grown both more numerous and more complex since the mid-1990s. Key drivers of the increasing demands include the complexity of service contracting, which is a growing share of all government contracting; the fact that the number of transactions is no longer a good measure of workload; and the fact that best-value procurement approaches are substantially more complex than lowest-price contracting approaches. The Section 1423 Report (2007, p. 19) concludes that

---

[2]   The sources calling for AW improvement acknowledge that workforce issues are only part of the problem. For example, in discussing the barriers to effective requirements determination, the Section 1423 Report (2007, p.7) not only points toward a strained workforce that lacks the requisite market expertise, but also to other factors that contribute to poor outcomes, such as a culture that emphasizes "getting to award," budgetary pressures, time pressures, and unclear roles and responsibilities—particularly in the use of interagency or government-wide contracts.

the demands placed on the acquisition workforce have outstripped its capacity. And while the current workforce has remained stable in the new millennium, there were substantial reductions in the 1990s accompanied with a lack of attention to providing the training necessary to those remaining to effectively operate the more complex buying climate.

2. DoD overuses or inappropriately uses contractors to perform acquisition functions. The dramatic increase in the federal government's use of contractors to provide services has received significant attention in recent years. Concerns relate not only to the number of contractors performing government functions, but also to the role they are playing—in particular, whether they are performing inherently governmental functions. Rostker (2008) argues that it is time for the federal government to rein in and rationalize the use of contractors.

Similar points have been made with respect to the defense AW. There is broad recognition in DoD that the contractor workforce has grown (Rostker, 2008; Section 814 Report, 2007), and congressional actions have prompted the Department of Homeland Security and DoD to review and reassess the way they are using service contractors (Rostker, 2008, p. 13).

3. The workforce lacks the skills to accomplish the workload. Another common refrain in discussions about the state of the defense AW is that the nature of the work has become substantially more complex, while the workforce has lost some of the skills or training needed to perform this work. This point is made in each of the reports discussed above. Increased workload complexity is attributed primarily to increased use of best-value procurement methods and the complexity of service contracts, which comprise a growing share of the workload. Evidence that the workforce lacks the skills necessary to fulfill its mission is largely anecdotal, and the arguments are far less specific than those related to workforce size.

DoD has announced plans to increase the defense AW by 20,000 (or 16 percent) over the next five years. The workforce plan has been described as a "bold step" toward addressing cost growth and schedule delays with major weapon systems (Hedgpeth, 2009). The proposed growth would include the conversion of 11,000 contractor support personnel to full-time government positions as well as 9,000 new federal hires.

It is unclear whether this step will, in fact, deliver on its promise of improving acquisition outcomes. Unfortunately, for all the information we have on acquisition outcomes and the AW, there is a dearth of evidence regarding whether and to what extent specific workforce issues are actually contributing to these outcomes. This paper assesses the evidence regarding the relationship between the issues described above and acquisition outcomes, and it discusses efforts that could inform future policy decisions related to the defense AW.

> **It is unclear that increasing the size of the acquisition workforce will improve acquisition outcomes.**

In the next section of this paper, we provide an overview of the defense AW and the policy environment influencing its management. In the third section, we assess the strength of the evidence supporting the key concerns that have emerged related to the AW. The final section offers conclusions and recommendations.

The Defense Acquisition Workforce: Policy Context, Size, and Composition

This section provides some critical background needed to understand the context for AW management and to assess the extent to which workforce issues may be affecting acquisition outcomes.[3] The management of federal government employees is subject to myriad external pressures and extensive oversight at various levels. The defense AW has received substantial additional attention over the years, making it,

---

[3]   This chapter draws heavily on material contained in Gates et al., 2008.

arguably, the most heavily scrutinized work-force in the federal government.

The federal AW includes men and women across all federal agencies who are responsible for acquiring the goods and services that their organizations need. The DoD portion of the federal AW, as defined by the official DoD AW count, consists of over 130,000 military and civilian employees, as well as a large number of contractors. The defense AW includes individuals

Defense acquisition personnel work across a wide variety of functional areas and organizations within the military services and defense agencies.

> responsible for planning, design, development, testing, contracting, production, introduction, acquisition logistics support, and disposal of systems, equipment, facilities, supplies, or services that are intended for use in, or support of, military missions. (DoD Instruction 5000.55, 2005)

Because of the breadth of the work carried out, the AW spans organizational boundaries within the Department of Defense to include the Army, Navy, Marine, Air Force, Defense Logistics Agency, and other entities within the Office of the Secretary of Defense (DoD Instruction 5000.55, 1991).

Defense Acquisition Workforce and Improvement Act

The policy environment for the management of the defense AW is dominated by the Defense Acquisition Workforce and Improvement Act (DAWIA) of 1990. DAWIA had its roots in DoD acquisition scandals of the mid-1980s[4] that led to internal and external pressures for reviews of defense acquisition processes, including President Reagan's Packard Commission. The consensus that emerged from these reviews was that

---

[4]   See Fairhall, 1987.

the defense AW underperformed and was too large. DAWIA attempted to address these size and quality concerns by requiring that DoD count and track the size of the AW and by imposing requirements on the training of acquisition workers, both military and civilian, employed by DoD.

DoD Instructions 5000.55 and 5000.66 are the key policy documents issued in response to DAWIA. Among other things, these instructions established twelve AW career fields;[5] provided guidance for managing the selection, placement, and career development of those filling positions within the AW; and defined workforce reporting requirements.

The military acquisition workforce is substantially smaller than the civilian acquisition workforce— but they have followed similar trends in terms of increases and decreases over time.

Since 1992, DoD has, consistent with DoD Instruction 5000.55, reported the number of military and civilian workers it employs who are part of the official AW (referred to as the "DAWIA count"). Figure 6.1 displays the civilian AW end-of-fiscal-year totals according to this DAWIA count. The figure shows that the civilian AW declined through the 1990s, reaching a low of 77,504 as of September 30, 1999. It then climbed steadily to 119,251 as of September 30, 2005, and then was reduced slightly to 113,605 by September 30, 2006.

The military AW is substantially smaller than the civilian workforce, but the trends have been consistent with those observed on the civilian side. The military AW stood at just over 16,500 in 1992; declined to 9,311 in 2000; and had increased to 14,976 by 2006.

---

5   The career fields are: Program Management; Communications-computer systems; Contracting; Purchasing; Industrial Property Management; Business, Cost Estimating, and Financial Management; Auditing; Quality Assurance; Manufacturing and Production; Acquisition Logistics; Systems Planning, Research, Development, and Engineering (SPRDE); Test and Evaluation Engineering. The Manufacturing and Production career field was eliminated in 2007 and a new career field, SPRDE Program Systems Engineer, was added in 2008.

Figure 6.1
Civilians in the Acquisition Workforce, September 30 Annual Snapshots

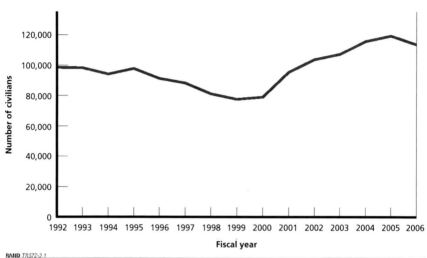

SOURCE: Gates et al., 2008, Figure 3.1.

As discussed in Gates et al. (2008, pp. 44–45), the services vary dramatically in terms of the size and composition of their organic AW; the Air Force employs the largest number and proportion of military personnel.

Figure 6.2 presents the career field distribution for the civilian AW. The majority of DoD civilian acquisition workers are employed in one of two career fields: (1) systems planning, research and development and engineering (SPRDE; 30 percent) or (2) contracting (22 percent). Only 7 percent of civilians are in program management. Figure 6.3 reveals a dramatically different career field distribution for military personnel.

Although contracting and SPRDE are important career fields for the military AW, the largest share of military acquisition workers is in the program management career field (29 percent of the total). Military personnel rarely fill positions in a number of fields, such as auditing,

Figure 6.2
Career Field Distribution for the Civilian Acquisition Workforce, FY 2006

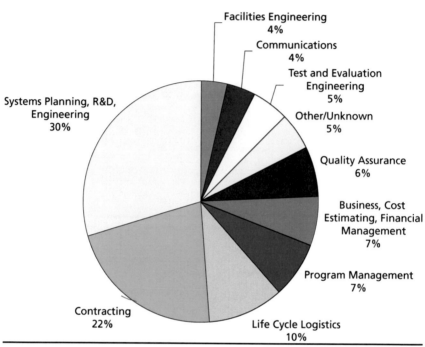

SOURCE: Gates et al., 2008, Figure 3.5.

science and technology, manufacturing and production, purchasing and procurement, and industrial property management.

Civilian Personnel Management in DoD: The National Security Personnel System

As illustrated in the previous section, civilian personnel dominate the organic AW. In 2003, Congress approved DoD's request to create a new human-resource management system for DoD's civilian workforce to replace the more traditional personnel management system. The National Security Personnel System (NSPS) is based on personnel demonstration projects that had been approved and implemented

Figure 6.3
Career Field Distribution for the Military Acquisition Workforce, FY 2006

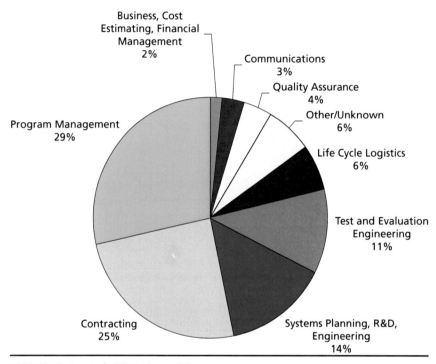

SOURCE: Gates et al., 2008, Figure 6.6.

since 1980 on a limited basis across the federal government. NSPS is intended to increase management flexibility in hiring, compensation and labor relations and to better motivate effective work. Importantly, NSPS allows DoD to link salary adjustments more directly with individual and organizational performance (Congressional Budget Office [CBO], 2008, pp. 22–23). DoD views flexibilities embodied in NSPS as critical to the effective recruitment and utilization of civilian personnel within DoD (CBO 2008, pp.1–2). The department began converting personnel to the NSPS system in 2006 and by the end of FY 2008,

26 percent of DoD's civilian workforce was part of NSPS.[6] Further expansion of NSPS has been put on temporary hold, pending a review of its implementation (DoD, 2009).

The AW had substantially more experience with the demonstration projects that inspired NSPS than has DoD's workforce as a whole. As of the end of FY 2005, 24 percent of DoD's AW was part of a demonstration project pay plan, compared with 7 percent of the overall DoD workforce.[7] The Acquisition Workforce Personnel Demonstration Project, as its name would suggest, focused specifically on the AW, and others, such as the demonstration project at the Naval Weapons Center in China Lake, California, focused on locations with a large share of acquisition workers. This suggests that the AW and its managers may be better prepared to implement NSPS and to reap the hoped-for benefits of the new system.[8]

> **Military personnel make up one-third of the program management community—but there are few military personnel in auditing, science and technology, and the manufacturing and production communities.**

Strategic Human Capital Planning for the Acquisition Workforce

The Department of Defense generates a DoD-wide strategic human capital plan for its entire civilian workforce.[9] The Under Secretary of Defense for Acquisition, Technology, and Logistics (USD (AT&L))

---

[6]   Author's calculations based on FY 2008 DoD Civilian Master File.

[7]   Author's calculations based on FY 2005 DoD Civilian Master File and Acquisition Workforce File data.

[8]   Evaluations of the demonstration projects provide a basis for encouragement, but no definitive evidence that the management flexibilities improve outcomes. For example, Schay et al. (2002) found that the demonstration project shifted employee expectations, albeit slowly, about the relationship between pay and performance.

[9]   The President's Management Agenda of 2001 emphasized the importance of improved management and performance of the federal government; a key initiative in the agenda is the effective strategic management of human capital within the government agencies.

issued a strategic human capital plan for the AW for the first time in 2006, which is currently in its third revision (see U.S. Department of Defense, Acquisition, Technology, and Logistics, 2007). This plan emphasizes the importance of a "total workforce" perspective that includes military, civilians and contractors. Subsequently, Section 851 of the National Defense Authorization Act of 2008 required DoD to have a separate section in its Civilian Human Capital Strategic Plan (HCSP) on the AW. The AW is the only workforce that has been singled out for special focus with a stand-alone, DoD-wide HCSP.

The acquisition workforce is the only workforce for which DoD has developed a stand-alone human capital strategic plan.

Although USD (AT&L) is the senior official providing overall supervision of the defense acquisition system, the office does not have direct authority over the many issues affecting AW management.[10] That authority falls to the services and agencies, which have "considerable influence over the shaping of their respective acquisition arms—prioritizing and approving operational requirements; building their [s]ervice program objective memorandums; and, in most cases, staffing and equipping program management offices" (Lumb, 2008, p. 19, summarizing findings from the Section 814 Report). USD (AT&L) is responsible for developing the AW plan and improving the AW; but ultimately, its role is to provide leadership and guidance on workforce issues.

The current emphasis on strategic human capital management is one of many workforce-related perspectives that have captured the attention of Congress and other federal policymakers over the years. We have already mentioned the pressures that emerged from the Packard Commission in the late 1980s to trim the size of the federal AW— the same workforce that is now criticized for being too small and for relying overmuch on contractors. These calls were buttressed by a more

---

[10]   See the Section 814 Report, 2007, pp. 2–5.

general movement to reinvent and downsize the federal government that began in 1993 with the National Performance Review. Following closely on the heels of that downsizing push was an emphasis on outsourcing or "contracting out" for goods and services in the mid-1990s. Managers may have shifted work from civilian and military personnel to contractor personnel in direct response. We cannot examine whether such a shift actually occurred because, as we discuss in the next section of this paper, we lack data on the use of contractors. In addition to these broader pressures to increase or decrease the use of certain types of personnel, special attention has also been paid at times to the staffing of particular types of organizations (such as DoD Headquarters organizations) or special types of personnel (such as senior executives or flag officers). Although strategic human capital management argues for organizing work and managing people in a strategic and effective way from a total workforce perspective, the current emphasis on this approach is just one of many pressures to which government managers are subject.

Basis of Main Concerns About the Defense Acquisition Workforce

In this section we describe the evidence in support of three critical issues that have been raised about the AW: that it is too small to meet current workload, that it lacks the skills needed to effectively accomplish the workload, and that the workforce mix is out of line in terms of the number of contractors being used to perform acquisition functions. We argue that the information available on workforce requirements, size, quality, and mix is insufficient to assess whether more workers, more highly skilled workers, or a different mix of workers would improve acquisition outcomes.

Is the Defense Acquisition Workforce Really Too Small?

To answer this question, one needs information about how many people are needed to accomplish the work (workforce demand) and how many people are currently part of the AW (workforce supply). No systematic data are currently available or referenced in workforce critiques on defense acquisition workforce demand. This is a key barrier to answering the question posed above since a characterization of the required workforce must anchor any assessment of whether the current workforce is too small or too large. Data on workforce supply exist, but they have serious limitations for accurately depicting trends in the size of the defense AW. Two limitations are of particular importance: (1) varying definitions of the organic (military and civilian) defense AW and (2) the absence of DoD-wide information on the number of contractors in the defense AW.

DoD data are based on definitions of the acquisition workforce that vary over time; we drew on data about DoD's overall civilian workforce for a more consistent view.

DoD recognizes that workforce management efforts must take a "total force perspective" that includes all military, civilian, and contractor personnel. A key barrier to the total force perspective for AW management is a lack of systematic data available on the contractor workforce (GAO, 2009b). Because information on the contractor workforce is completely lacking and because the military portion of the workforce is so small, discussions of AW size tend to focus on the organic, civilian workforce. Even there, data availability poses serious barriers to an analysis of the workforce.

For all the attention that has been focused on the defense AW over the past three decades, one would think there would be a clear and consistent definition of what the defense AW is, but this is not so. DoD has identified and gathered data on civilian and military members designated as part of the defense acquisition workforce (AW) since 1992. However, the definition used to identify these individuals has changed substantially over time—so much so that the Section 1423 Panel concluded that the data cannot be used to provide meaningful evidence of any personnel trends.[11]

---

[11] The Section 1423 Report provides a detailed discussion of the AW counting methods employed by DoD and by the federal government as a whole (p. 346–350).

Because DoD data are based on a definition of the AW that is not consistent over time, we performed an analysis of DoD-wide data to provide a new perspective on how changes in the size of the civilian defense workforce may be related to acquisition issues and compared our result to the official count of the defense AW.[12] Rather than focus on official defense AW data, we first looked at *DoD-wide civilian personnel* data, focusing on the number of DoD civilians in occupational series that are closely related to the acquisition activities described above. We also examined the number of DoD civilians in those occupational series who were counted as part of the official defense AW from 1992 to 2007.

> **Whereas the official DAWIA workforce has increased since 1992, we found that the number of DoD civilians in acquisition-related occupations had declined.**

Our analysis of DoD Civilian Personnel Master File data from the Defense Manpower Data Center, presented in Figure 6.4, shows that the total number of DoD civilians in key acquisition-related occupational groupings had increased through the 1980s, reached a peak in 1992, reached a low point in 2000, and has increased since then, but has not returned to 1992 levels (the 2007 level is 14 percent lower than in 1992). In contrast, between 1992 and 2007, the number of DoD AW civilians (as measured by the *official workforce count* in these same occupations) *increased* by 14 percent (see Figure 6.5). Thus, whereas trends based on the official AW count (depicted in Figure 6.1) reflect modest workforce growth since 1992, an analysis that is less tied to the arbitrary DAWIA (Defense Acquisition Workforce Improvement Act) workforce definition suggests a slight decline in the workforce over the same period of time.

---

[12] To perform this analysis, we used data that RAND assembled to support AW analysis in DoD. These data are described in detail in Gates et al., 2008. In identifying the occupational groupings for this analysis, we were guided by the Section 1423 Panel recommendations regarding which types of personnel should be considered part of the AW (Section 1423 Report, p. 344). We were also guided by FY 2007 DoD AW data. We attempted to identify occupational series for which designated members of the defense AW represent a large share of the overall DoD workforce. Details on the specific occupational series included in each grouping are described in the Appendix.

Figure 6.4
Number of DoD Civilians in Acquisition-Related Occupational Series (1980–2007), Drawn from Overall DoD Civilian Personnel Data

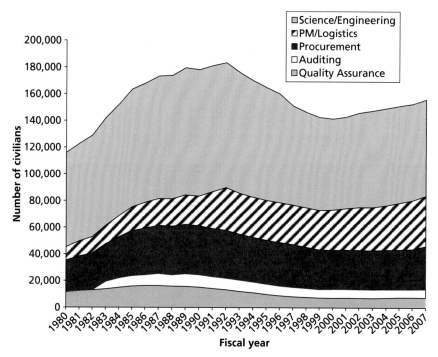

The dichotomy between trends based on the data we analyzed and trends based on official AW counts can be explained by shifts in the AW definition—in particular, an increased emphasis on including scientists and engineers in the DAWIA workforce count. Whereas in 1992, 38 percent of DoD personnel in acquisition-related engineering occupational series were counted as part of the AW, that figure was 65 percent by 2007. The implications are summarized in Figure 6, which depicts the number of all DoD civilians in science and engineering ("AW-related Occser Science/Engineering") versus the number of these who were counted as part of the official acquisition workforce ("AW Science/Engineering"); it also depicts the number of all DoD civilian in areas "other" than science and engineering ("AW-related Occser Other") versus the number of these who were counted as part of

Figure 6.5
Number of DoD Civilians in Acquisition-Related Occupational Series Classified as Part of the Official Acquisition Workforce Count (1992–2007)

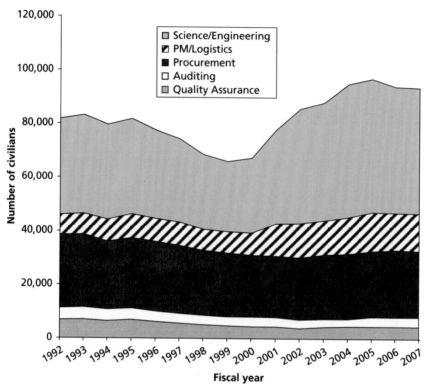

the official acquisition workforce ("AW Other"). Figure 6.6 illustrates that the number of DoD civilians in acquisition-related occupations increased dramatically between 1980 and 1992, began to decline until about 2001, and then experienced slight growth through 2007. In contrast, the number of individuals in these occupational series who were counted as part of the AW declined between 1992 and 2001 for scientists and engineers and then increased substantially after 2001. For other acquisition-related occupational series, the number counted as part of the AW was relatively stable between 1992 and 2007.

The modest growth in the official AW count also masks divergent trends by occupational series, which can be seen when the DoD-wide

Figure 6.6
Number of DoD Civilians in Science and Engineering and Other AW-Related
Occupational Series (Occser), Overall and in the Official Acquisition
Workforce Count (1980–2007)

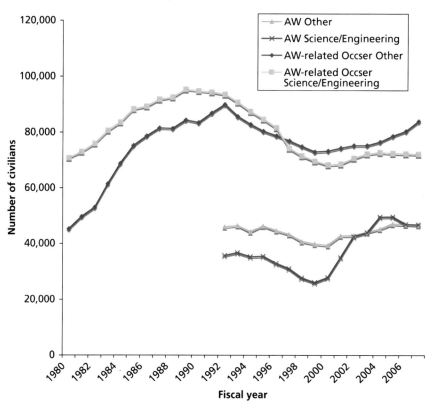

civilian workforce data are disaggregated. For example, the number of DoD civilians in the program management and logistics occupation series has increased substantially and consistently since 1980. In addition, the share of the DoD workforce in these occupational groupings counted as part of the official AW has increased from 1992 to 2007 (see Figure 6.7). In contrast, the total number of DoD civilians in the contracting, quality assurance, and auditing areas has declined steadily since the late 1980s. Figure 6.8 depicts the data for quality assurance. The share of the DoD civilian workforce counted as part of the official

Figure 6.7
Number of All DoD Program Management Civilians and Percentage Included in Official Acquisition Workforce Count

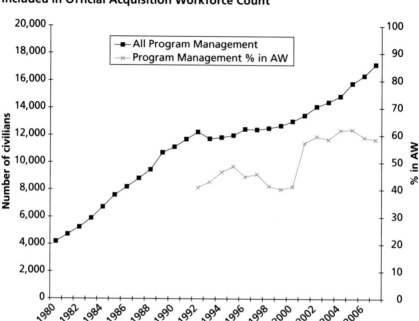

AW has been relatively stable since the official AW count began in 1992, but the *total* number of DoD civilians in these occupational areas has declined since then. The decline is most striking in quality assurance (44 percent) but is also substantial for auditing (26 percent) and for the more narrow contracting grouping (23 percent). The decline is 10 percent for the broader contracting grouping.

Contracting, quality assurance, and auditing— areas most likely to be affected by increased workload from procurement reforms—have experienced significant decreases in size.

This analysis suggests that trend analysis based on the official (DAWIA) AW count is misstating trends since 1992. Whereas official statistics suggest growth, there has likely been a slight decline in the size of the workforce. It also suggests that the contracting,

Figure 6.8
Number of All DoD Quality Assurance Civilians and Percentage
Included in Official Acquisition Workforce Count

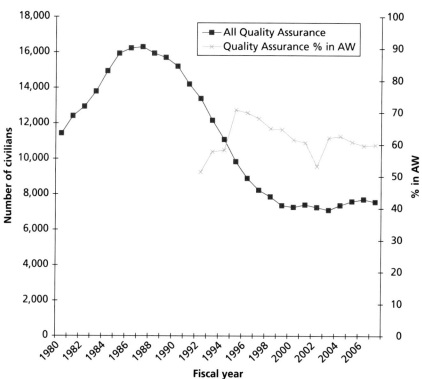

quality assurance, and auditing occupational groups—groups that would likely have been most affected by increased workload stemming from procurement reforms and increases in service contracts described above—have experienced the most significant declines in workforce size over time.

It is important to note that these data cover only the organic DoD civilian workforce; we do not know what role contractors are playing in these areas and cannot conclude anything about the growth or declines in *total* workforce size in these areas. As such, our analysis should be viewed as suggesting areas worthy of further examination rather than direct evidence that the workforce is too small in these areas.

Overuse or Inappropriate Use of Contractors

Due to lack of data, we are simply unable to characterize when, where, and why contractors are being used to provide acquisition-related services across DoD; the characteristics of those contractors; and how their use and characteristics may have changed over time. The information that we do have comes from targeted, in-depth, point-in-time examinations of specific programs or specific organizations. The major take-away from these studies is that DoD makes substantial use of contractors in performing acquisition-related functions and that this use varies dramatically across functions, occupations, programs, and organizations.

Where Are Contractors Being Used?

Targeted studies of the use of contractors to perform acquisition functions suggest that contractors are being used by most DoD acquisition organizations, but that organizational reliance on contractors is highly variable. Vernez et al. (2007) examined FY 2004 workforce data for individual business units within the Air Force Materiel Command (AFMC), which is the key acquisition command in the Air Force. Their analysis shows that the share of contractor manpower by organizational unit within AFMC varies dramatically, from 9 percent (Oklahoma Air Logistics Center) to as high as 89 percent (Arnold Engineering Development Center). Contractors represent 31 percent of the workforce at acquisition centers, 23 percent of the workforce at laboratory directorates, and 47 percent of the workforce at test and evaluation centers within AFMC. Other business units within AFMC, including logistics centers, have a lower reliance on contractors (Vernez et al., 2007, Table 2.2, p. 14). The authors also found substantial variation in the use of contractors

> Information on contractors is based on point-in-time studies of specific organizations; we cannot characterize when, where, and why contractors are being used to provide acquisition-related services across DoD.

across occupational areas; for example, 8 percent of engineers and scientists in the Aeronautical Systems Center were contractors, 55 percent in the Air Armament Center were contractors, and 78 percent in the Electronic Systems Center were contractors (Vernez et al., 2007, p. 13).

GAO's analysis of program office staffing for 61 major weapons programs showed that 41 percent of program office staff consisted of contractors. The largest number of contractors was found among engineering and technical staff, where 53 percent were not government employees. Twenty-six percent of staff in program management, 17 percent in contracting, and 47 percent in other business functions were contractors or other nongovernment staff. The fractions were substantially higher for administrative support and other areas (GAO, 2009a, p. 24).

Vernez and Massey (forthcoming), in research on the Air Force cost-estimating workforce, conducted a comprehensive point-in-time survey of all individuals working on cost-estimating tasks for the Air Force in spring of 2008. Their work reveals that about half of the individuals performing such tasks are contractors and that the proportion of contractors in the workforce varies across the product and logistics centers examined. Contractors did not appear to be any less qualified; they were about as likely to have certification in the area of cost estimation (about one-third of the workforce had such certification) and tended to have as much or more experience in cost estimation compared with the organic workforce. The study also found that the Air Force was relying on contractors to do the actual cost-estimating work, whereas the organic staff tended to be financial management generalists or cost managers in charge of managing the cost-estimating work and integrating that work with other financial management functions (Vernez and Massey, forthcoming).

In a 2006 DoD Inspector General (DoD IG) audit of the AW at six AW locations, one location (Naval Sea Systems Command) was unable to provide any data on contractors performing acquisition functions. At the other locations, contractors as a portion of the total AW ranged from 16 percent (Defense Supply Center Columbus) to 64 percent (Air Force Space and Missile Systems Center) (U.S. Department of Defense, Office of the Inspector General, 2006, p. 12).

Why Are Contractors Being Used?

The targeted studies described previously suggest that the use of contractors to perform acquisition functions is not based on a comprehensive strategic assessment of the long-run costs and benefits of their use. The studies also provide evidence that acquisition organizations are not able to fill all of their requirements with in-house personnel for a range of reasons, including resource constraints and process barriers.

> It is not clear that DoD's use of contractors to perform acquisition functions is based on a strategic assessment.

In its review of staffing for major weapons programs, GAO found that 46 out of 59 programs that responded to questions about staffing had received authorizations for all required positions, but that only 42 percent of the programs were able to fill all of their authorized positions (GAO, 2009a, pp. 23–24). Thirty-one of these programs provided information to GAO about the reasons for using contractor personnel (GAO, 2009b, pp. 8–9). Only one cited cost considerations. Over three-quarters of DoD acquisition programs reported that they used contractors as a way to get around critical constraints: personnel ceiling, civilian pay budget constraints, limitations with the federal government hiring process, or a lack of in-house capability in a particular area. These findings echoed those from a prior study conducted by the DoD Inspector General (DoD IG, 2006, p. 13).

GAO attributes DoD's reliance on contractor support to a "critical shortage of certain acquisition professionals with technical skills as it [DoD] has downsized its workforce over the last decade" (GAO, 2008, p. 30). GAO's report also noted that some of the program offices interviewed for its study expressed concerns about inadequate manpower. GAO found that DoD has given contractors increased responsibility for "key aspects of setting and executing a program's business case," including requirements development and product design (GAO, 2008, p. 29).

Vernez et al. (2007, p. 13) found varying perspectives on the pros and cons of using contractors to perform acquisition functions:

When line managers in the AAC were asked what the optimum share of contractors would be, their answers ranged broadly, from a low of 20 to a high of 80 percent. The low figure reflected respondents' concerns for continuity and institutional memory. The high figure reflected the view that contractors could do most of the functions of an SPO [system program office] with the exceptions of the director and key financial, security, and contracting positions.

The authors suggest that there was no way to assess the efficiency of the use of contractors in the organizations they studied.

Vernez and Massey (2009) found that the actual workforce in the cost-estimating area was about 75 percent of stated requirements. Those interviewed for the study pointed to challenges in filling positions as well as a failure to obtain hiring authorizations for all the requirements as reasons why the actual workforce fell short of requirements.

The Workforce Lacks the Skills to Accomplish the Workload

Another common refrain in discussions about the state of the defense AW is that the nature of the work has become substantially more complex, while the workforce has lost some of the skills or training needed to perform this work. The drivers of increased complexity were discussed earlier and are primarily attributed to increased use of best-value procurement methods and the complexity of service contracts.

Evidence that the workforce lacks the skills necessary is largely anecdotal, and the arguments are far less specific than those related to workforce size. A key barrier to assessing this perspective is a lack of systematic data on the skill level of the workforce, not to mention the skills that are required to perform the work (GAO, 2009b, p. 9). The only data available on the AW that are remotely related to workforce quality are certification levels and education levels. This information is available only for the organic workforce.

On the basis of available information, the situation looks pretty good. According to the Section 814 Report (p. 3-2), the AW is more experienced and more highly educated than the defense workforce overall, and certification rates are high:

> 66 percent of the AT&L civilian workforce is certified, and 50 percent meet or exceed the required position certification level. However, for critical acquisition positions, the certification rate increases to 75 percent, with 65 percent meeting or exceeding the position.

But it is not clear how well certification standards—and the training provided to achieve those standards—are aligned to actual skill requirements. Moreover, concerns have been expressed that even the certification standards may be outdated or that the training provided to meet the standards does not reflect current skill needs (Vernez and Massey, forthcoming).

DoD lacks systematic data on the skill level of the workforce—and it is not clear how existing data, such as certification standards, are aligned to actual skill requirements.

USD (AT&L) is leading an effort to define workforce competencies for critical segments of the AW (GAO, 2009b, pp. 10–11). This effort should lay the groundwork for a more systematic analysis of the question of whether the workforce actually has the skills needed to do the work. A big question that must be addressed in any analysis of this issue is whether observed deficits in skills stem from a lack of training, from an inappropriate workforce mix, or from a combination of both.

Conclusions

The AW has been the subject of numerous investigations and specific policy guidance over the past three decades. There have been pressures to increase and decrease the size of this workforce, to improve its quality (usually in terms of training and certification requirements), and to both outsource its workload and bring its workload in-house. Yet few would argue that defense acquisition outcomes have dramatically improved in response to these varied policy initiatives.

The FY 2010 defense budget proposal includes the latest installment in a series of policy initiatives targeting the defense AW—with recommendations to grow the workforce and rein in reliance on contractors. But should we expect that a larger federal defense AW will lead to improved acquisition outcomes? Workforce initiatives are unlikely to be the silver bullet that will improve acquisition outcomes, but given present data constraints, we would not be able to answer that question anyway. As we have demonstrated in this paper, the information needed to assess the success of workforce initiatives and their contribution to overall acquisition outcomes is sorely lacking.

> **Workforce initiatives are unlikely to be the silver bullet that will improve acquisition outcomes.**

Efforts should be directed toward assembling the information needed to track the effectiveness of these new initiatives and to make, refine, or dismiss the case for further workforce adjustments. Below, we identify the steps DoD should take to acquire this information.

Establish Key Process Standards That Are Plausibly Influenced by the Workforce, and Consistently Monitor Those Processes. An infinitely large and supremely qualified AW will not generate on-time, on-budget systems with no problems or appeals 100 percent of the time. The AW acts within the confines of a process, and if the process itself is not operating effectively, then improvements to the workforce can only do so much. Attention must be paid to the acquisition process itself, including the incentives for effective work embodied in that process.

The AW must be viewed as an input to a process operation, and thought should be given to concrete outcomes that the workforce could be expected to influence. These would not be the high-level outputs of on-time, on-budget systems, but they could include important process-oriented outcomes that reflect top-flight systems engineering practices and could ultimately lead to improvements in the key outcomes of interest. It is also critical to acknowledge that the AW is engaged in a wide range of procurement-related activities and that different types of activities are likely to require separate and distinct outcome measures.

Map Workforce Characteristics to Acquisition Activities and Their Outcomes. To identify the impact of workforce size and quality on acquisition outcomes, one needs to assess acquisition outcomes and relate those outcomes back to the workforce. For traditional defense acquisition systems, it may be sensible to track data at the acquisition program level using data on the workforce that are mapped to acquisition programs. An ability to map the AW to outcome data for the programs or organizations in which they work would support systematic analyses of the relationship between workforce attributes and outcomes. Currently, such a mapping of the defense AW is not possible

Accomplishing this goal would require managers to develop metrics appropriate to the program, organization, or activity in question that plausibly inform the quality of the work being done; that is, they should develop metrics based on the things that the workforce could influence and that would ultimately be expected to affect outcomes. An improved ability to link the workforce with organizational outcomes is consistent with strategic human capital management and with an effective implementation of NSPS. For example, if managers agree that providing timely systems engineering to support investment decisionmaking is a critical process indicator, they could track whether such activities are occurring and possibly assess the quality of those activities. That information could then be linked with data on that program's workforce to assess the relationship between workforce characteristics and these outcomes. Similarly, the tenure of program managers has been highlighted as a plausible factor influencing outcomes (GAO, 2008, p. 29). This workforce characteristic could be tracked at the program level and related to program outcomes to determine whether there exists a relationship between tenure and outcomes.

Assess the Appropriateness of the Current Workforce Mix. As illustrated in this paper, the data required to provide a convincing argument that the defense AW mix is inadequate or inappropriate to meet current needs are lacking. Our analysis reveals declines in the number of DoD employees in auditing, contracting, and quality assurance occupations. Contractors may have been used to fill some of this gap. Our analysis also suggests that the AW focused on science and engineering has remained relatively stable and that program managers and logistics professionals (generalists) have grown. But current data cannot shed light on whether the workforce mix is appropriate and adequate to workforce needs.

The ideal workforce mix is likely to vary by acquisition activity and to change over time as acquisition processes and priorities change. Assessing whether the workforce mix is on target requires data that relates workforce measures to outcomes using a consistent unit of analysis such as the acquisition program. Because it will take time to assemble such data and identify the critical process and outcome data, it may be worthwhile for DoD to conduct a rough assessment of the appropriateness of the workforce mix through a systematic, program-by-program survey of program managers. Such information could be rolled up to provide a rough, high-level sense of some critical areas where the workforce mix is out of balance and to suggest more short-term actions that might be taken to correct some imbalances.

Include the Contractor Workforce in Strategic Workforce Planning. Currently, contractors are effectively ignored in strategic human capital efforts, yet we know they are playing a nontrivial role. The bottom line is that it is not possible to effectively manage human capital while ignoring an important segment of the workforce. In order to better understand the use of contractors in acquisitions, two things are needed: (1) better data on the contractor workforce as discussed above, and (2) a better understanding of the environment in which acquisition-related staffing and resource decisions are made.

Assess How Staffing and Resourcing Decisions Related to Acquisition Functions Are Made. Policymakers must keep in mind that specific characteristics of the workforce and its training and development are only partial contributors to acquisition outcomes. Even policies that are targeted specifically at the AW are influenced by budgeting and management decisions that take place within the services and agencies. A realistic assessment of how staffing and resource decisions relate to the acquisition functions—the decisions that determine how many and what type of people are brought onboard to do the work, how their workload is managed, and how they are a mentored and trained—is necessary to understand the effect that specific policies are likely to have on the AW and ultimately on acquisition outcomes. Such an understanding is especially critical in a time of workforce growth because the hiring that takes place today will influence the AW for decades to come.

Occupational Grouping Definitions Used in This Report

To identify occupational groupings for this analysis, we were guided by the recommendations contained in the Section 1423 Report as to who should be considered part of the AW. We were also guided by FY 2007 DoD AW data. We attempted to identify occupational series for which designated members of the defense acquisition workforce (AW) represent a large share of the DoD workforce as a whole.

We considered the following occupational groupings:

- Quality Assurance
- Auditing
- Program Management and Logistics
- Procurement
- Science and Engineering

Quality Assurance and Auditing

In analyzing trends for quality assurance and auditing, we look at occupational series with 1910 (Quality Assurance) and 511 (Auditing).

Program Management and Logistics

In the program management and logistics area we provide two different slices on the data. A comprehensive program management and logistics category includes all the following occupational series: 340 (Program Management), 343 (Management and Program Analysis), and 346 (Logistics Management). The more narrow program management category includes 340 and 343. Note that we exclude occupational series 301 (Miscellaneous Administrative and Program) from both analyses. Although this occupational series represents a substantial share (21 percent) of civilians in the DoD AW Program Management career field, the AW represents only 13 percent of all DoD civilians in that occupational series in FY 2007. Trends are similar for the two groupings.

"Series 301, covers positions the duties of which are to perform, supervise, or manage two-grade interval administrative or program work for which no other series is appropriate. The work requires analytical ability, judgment, discretion, and knowledge of a substantial body of administrative or program principles, concepts, policies, and objectives." (U.S. Office of Personnel Management, 2008)

Procurement

In the procurement area, we also present two slices of the data. A more comprehensive grouping includes the following occupational series: 1101 (General Business and Industry), 1102 (Contracting), 1103 (Industrial Property Management), 1104 (Property Disposal), 1105 (Purchasing), and 1150 (Industrial Specialist).[13]

A more restricted grouping (Contracting 2) includes the following occupational series: 1102, 1103, 1105, 1150. The second grouping emphasizes those occupational series for which the AW is a 90 percent + share of the AW. Note that there are a large number of individuals in the 1101 series who are part of the AW, (3,816 in FY 2007), but they represent only 37 percent of the total DoD 1101 workforce. The overall trends are similar for both definitions.

Science and Engineering

To determine the acquisition-related science and engineering positions, we looked at the percentage of the DoD workforce that is

---

[13]   Note that some FAI analyses have included occupational series 1106, but there were fewer than 100 such individuals in the AW in 2007, and we have excluded them from the analysis.

counted as part of the AW by occupational series for the 800, 1300, 1500 series. We included all occupational series where the AW share was more than one-third in FY 2007. These are as follows:

801 General Engineering
803 Safety Engineering
804 Fire Protection Engineering
806 Materials Engineering
810 Civil Engineering
819 Environmental Engineering
830 Mechanical Engineering
850 Electrical Engineering
854 Computer Engineering
855 Electronics Engineering
858 Biomedical Engineering
861 Aerospace Engineering
871 Naval Architecture
890 Architectural Engineering
893 Chemical Engineering
896 Industrial Engineering
1301 General Physical Sciences
1310 Physics
1315 Hydrology
1320 Chemistry
1321 Metallurgy
1350 Geology
1370 Cartography
1382 Food Technology
1384 Textile Technology
1515 Operations Research
1520 Mathematics
1529 Mathematical Statistics
1550 Computer Science

The largest occupational series included in this analysis are: 801 (8,013 AW members; 70 percent), 810 (4,126 AW members; 65 percent), 830

(6,287 AW members; 69 percent), 854 (2,421 AW members; 81 percent), 855 (11,132 AW members; 67 percent), 861 (2,995 AW members; 82 percent), 1550 (2,564 AW members; 60 percent).

Although, as we report below, the share of DoD's science and engineering workforce that is counted as part of the AW has grown over time, we did not observe shifts in the specific occupational series that were included in the defense AW.

Generally speaking, those occupational series in the 800, 1300, and 1500 series that were excluded on the basis of this cutoff had 25 or fewer DoD AW members in FY 2007. The three exceptions are occupational series 802 (Engineering Technical), which had 1045 AW members representing 9 percent of the DoD workforce; occupational series 808 (Architecture), which had 255 AW members representing 32 percent of the AW; and occupational series 856 (Electronics Technical), which had 351 AW members representing 6 percent of the DoD workforce.

Sensitivity Checks

To validate that the occupational series considered part of the AW had not changed much between 1992 and 2007, we reviewed data on the percentage of the DoD workforce counted as part of the AW by occupational series for FY 1992. With a few exceptions (809, 856, 895, 1340, 1386, 1521, 1531), the share of the occupational series classified as part of the AW rose between 1992 and 2007. Among those occupational series where the share declined between 1992 and 2007, the share of all occupational series members in the AW was well below the one-third threshold in both years except in the case of occupational series 1386 (Photographic Technology), where it was 34 percent in FY 1992. However, this is a small career field (with only 26 members DoD-wide in FY 1992 and seven in FY 2007) that does not seem to be directly related to the acquisition; hence, we decided not to include it in the analysis.

About the Authors

Mark V. Arena has worked on a variety of research projects for the U.S. Navy, the U.S. Air Force, OSD, and the U.K. Ministry of Defence since joining RAND in 1998. Currently, he is the associate director for the Acquisition and Technology Policy Center in RAND's National Security Research Division. His research has focused on technology evaluations, cost analysis, risk analysis, and industrial-base simulations for acquisition programs. Recent projects includes a study of the long-term price changes for ships and fixed-wing aircraft, the history of cost growth for major weapon systems, the cost savings due to multiyear procurement for the F-22A, and an assessment of domestic submarine-design capabilities. He has also developed an analytical model for the U.S. Navy and the U.K. Ministry of Defence to understand the shipyard and budget impacts of various force-structure choices and procurement plans. He graduated from Yale University with a B.S. in chemistry and later received a Ph.D. in physical chemistry from Stanford University.

John Birkler has held a wide variety of research and management positions since joining RAND in 1977. Currently, he manages RAND's Maritime Program, overseeing research for the U.S. Navy, the Office of the Secretary of Defense, the U.S. Special Operations Command (SOCOM), the U.S. Coast Guard, the Australian Department of Defence, and the U.K. Ministry of Defence. He also mentors U.S. Navy, Marine Corps, and Coast Guard Executive Fellows at RAND. His research spans RDT&E strategies and planning and industrial base, acquisition, management, and organizational issues. In addition to the above maritime clients, his research has covered a wide range of aircraft systems (including the Joint Strike Fighter, the F-15, the F-14, the B-1, the B-2, the A-12, the C-5, the C-17, the F-117, the F/A-18 E/F), missiles and munitions (including the advanced cruise missile, the Tomahawk cruise missile, and precision conventional munitions), and surface and subsurface combatants. He also has led studies on the links between the health of the defense industrial base and levels of innovation and competition. His most recent work has involved managing or leading multiple analyses of alternatives for the U.S. Navy, the Marine Corps, the U.S. Army, and SOCOM and leading a high-profile RAND analysis of Australia's capabilities and capacity to design conventional submarines. He holds B.S. and M.S. degrees in physics and completed the UCLA executive program in management in 1992. After completing his third command tour, he retired from the Navy Reserve with the rank of captain.

Irv Blickstein is a senior engineer at RAND with 35 years of experience in the field of defense analysis and management; his specialties are planning, programming, budgeting, and acquisition. Since joining RAND in 2001, he has managed research activities associated with a series of projects in support of the office of the Chief of Naval Operations and the Undersecretary of Defense (AT&L), and he also serves on the Chief of Naval Operations Executive Panel. His projects have covered a wide variety of topics, including directed energy, unmanned vehicles, reviews of foreign acquisition programs, sea basing, and cost analyses of both ships and aircraft. He also has examined the cost of statutory and regulatory constraints in DoD. Recent research includes an analysis of cost growth in U.S. Navy ships over a 40-year period and an evaluation of the hiring and budgeting practices at naval shipyards. He served 31 years in DoD before joining RAND, 18 of them as a senior executive, and he was awarded four Presidential Rank awards. He received a B.S. in industrial engineering from Ohio State University and an M.E.A. in engineering management from George Washington University.

Jeffrey A. Drezner, a senior policy researcher, joined RAND in 1985. He has conducted policy analysis on a wide range of issues, including energy research and development planning and program management; best practices in environmental management; analyses of cost and schedule outcomes in complex system development programs; aerospace industrial policy; defense acquisition policy; and local emergency response. His current research continues to emphasize issues associated with technology development, organizational behavior, and program management. He served as the associate program director for acquisition in RAND's Project Air Force from 1992 to 1994. From June 1995 to September 1995, he participated in a researcher exchange program in the Pentagon, sponsored by the Office of the Under Secretary of Defense (AT&L). He received an M.S. in policy analysis from the University of California, Davis, and a Ph.D. in political science (public policy) from the Claremont Graduate School.

Susan M. Gates is a senior economist, the director of the Kauffman-RAND Institute for Entrepreneurship Public Policy in the RAND Institute for Civil Justice (ICJ), the quality assurance manager for ICJ, and a professor of economics at the Pardee RAND Graduate School. She specializes in the economics of organizations, political economy, and applications of economic management principles to public sector organizations. She has a special interest in organizational governance, entrepreneurship, leadership, and workforce management. She has led numerous projects on a wide range of topics over the years. Her current projects include a comprehensive evaluation of the New Leaders for New Schools national principal training and support program, data-driven analyses of the DoD acquisition workforce, and a study of the potential role of innovation in improving health care delivery. Previous studies examined organizational approaches for ensuring the quality and productivity of DoD's education and professional development activities targeting civilians, outsourcing in DoD, higher education from an industry perspective, and the cost of and demand for a military childcare program. She received her Ph.D. in economics from Stanford Graduate School of Business and her M.A. in Russian and Eastern European studies from Stanford University.

Meilinda Huang was a research assistant at RAND from 2007 to 2009. In addition to national security projects, she contributed to analyses related to health economics and RAND's Institute for Civil Justice. She worked with a large research team to develop a comprehensive framework for assessing how to evaluate the Medicare Part D program, both on its own and in relation to other large public and private prescription drug programs. In addition, she contributed to case studies on cell phone radiation, deca-BDE, and benzene, which pose significant health risks and could trigger the next mass litigation case. She also created a database that organized the compensation given to individuals and businesses by government agencies and programs, insurance, and charities in the aftermath of Hurricane Katrina. She received her B.A. in economics from Princeton University.

Robert Murphy retired as a member of the Senior Executive Service after a 33-year career in the U.S. Navy's Nuclear Propulsion Program. He finished his career as the program's director of resource management and was responsible for budget, acquisition, and logistic support for the Naval Nuclear Propulsion Program. While in government service, he participated in the acquisition of every nuclear submarine in the U.S. fleet today in addition to design contracts for *Seawolf-* and *Virginia*-class submarines. Since retiring from public service, he has been consulting for both commercial and government organizations, specializing in major system acquisitions. Projects have included several cost studies of Virginia-class submarines and advising on the most recent multiyear contract for acquisition of these submarines. RAND research projects have included several studies of the United Kingdom's nuclear submarine industrial base and of Australian conventional submarine design issues. He earned an M.B.A. from George Washington University.

Charles Nemfakos was appointed a senior fellow at RAND after leading Nemfakos Partners LLC in providing support to private sector and governmental entities, both in the United States and internationally. He provides research, strategically oriented analyses, support, and advice to a broad variety of RAND clients. Previously, he was an executive with

Lockheed Martin Corporation, Naval Electronics and Surveillance Systems. During his federal service, he served in assignments in the executive and legislative branches of government, primarily in DoD, as a budget analyst and as a planner in the Office of the Secretary of Defense and the Department of the Navy. As a senior executive in the Department of the Navy, he served in a variety of financial positions, including deputy assistant secretary for installations and logistics, deputy under secretary, and senior civilian official for financial management and comptroller. He was the recipient of multiple Distinguished Service Medals and numerous other awards and citations, and he has been recognized by three U.S. presidents with four Presidential Rank Awards. He was selected to receive American University's Roger W. Jones Award for Executive Leadership, elected a fellow of the National Academy of Public Administration, and honored by the Secretary of Defense as one of nine Career Civilian Exemplars in the 228-year history of the U.S. armed forces. He led two Department of the Navy operating units that were awarded the Meritorious Unit Citation. He received a B.A. in history from the University of Texas and an M.A. in government from Georgetown University.

Susan Woodward is a senior communications analyst, working principally for RAND's national security divisions to develop all forms of research communication and help ensure RAND that research is accessible to diverse audiences. Prior to joining RAND, she was a senior analyst in the U.S. Government Accountability Office's Homeland Security and Defense divisions, assessing the effectiveness of government programs related to aviation security, defense acquisition, and defense personnel issues. She received a B.A. in English from the University of Maine and an M.A. and Ph.D. candidacy in English from the University of Michigan.

Bibliography

Acquisition Advisory Panel, Section 1423 Report, *Report of the Acquisition Advisory Panel to the Office of Federal Procurement Policy and the United States Congress*, Washington, D.C.: Office of Federal Procurement Policy, 2007.

Antón, Philip S., Eugene C. Gritton, Richard Mesic, and Paul Steinberg, *Wind Tunnel and Propulsion Test Facilities: An Assessment of NASA's Capabilities to Serve National Needs*, Santa Monica, Calif.: RAND Corporation, MG-178-NASA/OSD, 2004. As of November 16, 2009: http://www.rand.org/pubs/monographs/MG178/

Antón, Philip S., Dana J. Johnson, Michael Block, Michael Scott Brown, Jeffrey A. Drezner, J. A. Dryden, E. C. Gritton, Thomas Hamilton, Thor Hogan, Richard Mesic, Deborah J. Peetz, Raj Raman, Paul Steinberg, Joe Strong, and William Trimble, *Wind Tunnel and Propulsion Test Facilities: Supporting Analyses to an Assessment of NASA's Capabilities to Serve National Needs*, Santa Monica, Calif.: RAND Corporation, TR-134-NASA/OSD, 2004. As of November 16, 2009: http://www.rand.org/pubs/technical_reports/TR134/

Archibald, Kathleen A., Alvin J. Harman, M. Hesse, John R. Hiller, and Giles K. Smith, *Factors Affecting the Use of Competition in Weapon System Acquisition*, Santa Monica, Calif.: RAND Corporation, R-2706-DRE, February 1981. As of November 16, 2009: http://www.rand.org/pubs/reports/R2706/

Arena, Mark V., Robert S. Leonard, Sheila E. Murray, and Obaid Younossi, *Historical Cost Growth of Completed Weapon System Programs*, Santa Monica, Calif.: RAND Corporation, TR-343-AF, 2006. As of November 16, 2009: http://www.rand.org/pubs/technical_reports/TR343/

Arena, Mark V., Lionel A. Galway, John C. Graser, Felicia Wu, Obaid Younossi, Bernard Fox, Jerry M. Sollinger, and Carolyn Wong, *Impossible
Certainty: Cost Risk Analysis for Air Force Systems*, Santa Monica, Calif.:
RAND Corporation, MG-415-AF, 2006. As of November 16, 2009:
http://www.rand.org/pubs/monographs/MG415/

Birkler, John L., et al., *Differences Between Military and Commercial Shipbuilding: Implications for the United Kingdom's Ministry of Defence*, Santa Monica, Calif.: RAND Corporation, MG-236-MOD, 2005.
http://www.rand.org/pubs/monographs/MG236/

Birkler, John L., Anthony G. Bower, Jeffrey A. Drezner, Gordon T. Lee, Mark
A. Lorell, Giles K. Smith, Fred Timson, William P. G. Trimble, and Obaid
Younossi, *Competition and Innovation in the U.S. Fixed-Wing Military
Aircraft Industry*, Santa Monica, Calif.: RAND Corporation, MR-1656-
OSD, 2003. As of November 16, 2009:
http://www.rand.org/pubs/monograph_reports/MR1656/

Birkler, John L., Edmund Dews, and Joseph P. Large, *Issues Associated with
Second-Source Procurement Decisions*, Santa Monica, Calif.: RAND
Corporation, R-3996-RC, 1990. As of November 16, 2009:
http://www.rand.org/pubs/reports/R3996/

Birkler, John L., John C. Graser, Mark V. Arena, Cynthia R. Cook, Gordon
T. Lee, Mark A. Lorell, Giles K. Smith, Fred Timson, Obaid Younossi,
and Jon Grossman, *Assessing Competitive Strategies for the Joint Strike
Fighter: Opportunities and Options*, Santa Monica, Calif.: RAND Corporation, MR-1362-OSD/JSF, 2001. As of November 16, 2009:
http://www.rand.org/pubs/monograph_reports/MR1362/

Birkler, John L., and J. P. Large, *Dual-Source Procurement in the Tomahawk
Program*, Santa Monica, Calif.: RAND Corporation, R-3867-DR&E,
1990. As of November 16, 2009:
http://www.rand.org/pubs/reports/R3867/

Birkler, John L., John F. Schank, Mark V. Arena, Giles K. Smith, and Gordon
T. Lee, *The Royal Navy's New-Generation Type 45 Destroyer: Acquisition
Options and Implications*, Santa Monica, Calif.: RAND Corporation,
MR-1486-MOD, 2002. As of November 16, 2009:
http://www.rand.org/pubs/monograph_reports/MR1486/

Birkler, John L., Giles K. Smith, Glenn A. Kent, and Robert V. Johnson,
An Acquisition Strategy, Process, and Organization for Innovative Systems,

Santa Monica, Calif.: RAND Corporation, MR-1098-OSD, 2000. As of November 16, 2009:
http://www.rand.org/pubs/monograph_reports/MR1098/

Bolten, J. G., Robert S. Leonard, Mark V. Arena, Obaid Younossi, and Jerry M. Sollinger, *Sources of Weapon System Cost Growth: Analysis of 35 Major Defense Acquisition Programs*, Santa Monica, Calif.: RAND Corporation, MG-670-AF, 2008.
http://www.rand.org/pubs/monographs/MG670/

Casey, John P., Electric Boat Corporation, testimony before the Seapower and Expeditionary Forces Subcommittee of the House Armed Services Committee, Washington, D.C., March 8, 2007.

Chadwick, Stephen Howard, *Defense Acquisition: Overview, Issues, and Options for Congress*, Washington, D.C.: Congressional Research Service, RL34026, June 20, 2007.

Chu, David, and Jeffrey Eanes, *The Trajectory of Personnel Costs in the Department of Defense*, Washington, D.C.: Office of the Under Secretary of Defense for Personnel and Readiness, 2008.

Commission on Army Acquisition and Program Management in Expeditionary Operations, "Gansler Commission Report," *Urgent Reform Required: Army Expeditionary Contracting*, October 31, 2007. As of December 12, 2008:
http://www.army.mil/docs/Gansler_Commission_Report_Final_07 1031.pdf

Comptroller General of the United States, Decision B-275830, April 7, 1997.

Congressional Budget Office, *A Review of The Department of Defense's National Security Personnel System*, Washington, D.C., November 2008.

Congressional Research Service, *Defense Acquisition Workforce: Issues for Congress*, Washington, D.C., March 11, 1999.

———, *Navy LPD-17 Amphibious Ship Procurement: Background, Issues, and Options for Congress*, updated May 23, 2008.

Defense Acquisition University, *Acquisition, Technology, and Logistics Knowledge Sharing System*, 2008. As of January 28, 2009:
http://akss.dau.mil

———, "Systems Acquisition Management: An Introduction," Lesson 2. As of September 15, 2009:
https://learn.dau.mil/CourseWare/1_13/media/Lesson_2_Summaryf.pdf

————, Section 814 Report, *Defense Acquisition Structures and Capabilities Review Report, Fiscal Year 2006*, Fort Belvoir, Va., June 2007.

DoD—*See* U.S. Department of Defense.

Drezner, Jeffrey A., *The Nature and Role of Prototyping in Weapon System Development*, Santa Monica, Calif.: RAND Corporation, R-4161-ACQ, 1992. As of November 16, 2009:
http://www.rand.org/pubs/reports/R4161/

Drezner, Jeffrey A., J. M Jarvaise, R. W. Hess, P. G. Hough, and D. Norton, *An Analysis of Weapon System Cost Growth*, Santa Monica, Calif.: RAND Corporation, MR-291-AF, 1993. As of November 16, 2009:
http://www.rand.org/pubs/monograph_reports/MR291/

Drezner, Jeffrey A., and Giles K. Smith, *An Analysis of Weapon System Acquisition Schedules*, Santa Monica, Calif.: RAND Corporation, R-3937-ACQ, December 1990. As of November 16, 2009:
http://www.rand.org/pubs/reports/R3937/

Drezner, Jeffrey A., and Robert S. Leonard, *Innovative Development: Global Hawk and DarkStar—Flight Test in the HAE UAV ACTD Program*, MR-1475-AF, Santa Monica: RAND, 2001a. As of November 16, 2009:
http://www.rand.org/pubs/monograph_reports/MR1475/

————, *Innovative Development: Global Hawk and DarkStar—HAE UAV ACTD Program Description and Comparative Analysis*, MR-1474-AF, Santa Monica: RAND, 2001b. As of November 16, 2009:
http://www.rand.org/pubs/monograph_reports/MR1474/

————, *Innovative Development: Global Hawk and DarkStar—Transitions Within and Out of the HAE UAV ACTD Program*, MR-1476-AF, Santa Monica: RAND, 2001c. As of November 16, 2009:
http://www.rand.org/pubs/monograph_reports/MR1476/

————, *Global Hawk and DarkStar: Transitions Within and Out of the HAE UAV ACTD Program*, Santa Monica, Calif.: RAND Corporation, MR-1473; MR-1474; MR-1475; MR-1476, 2002a. As of November 16, 2009:
http://www.rand.org/pubs/monograph_reports/MR1476/

————, *Innovative Development: Global Hawk and DarkStar—Their Advanced Concept Technology Demonstrator Program Experience, Executive Summary*, MR-1473-AF, Santa Monica: RAND, 2002b. As of November 16, 2009:
http://www.rand.org/pubs/monograph_reports/MR1473/

Drezner, Jeffrey A., Geoffrey Sommer, and Robert S. Leonard, *Innovative Management in the DARPA High Altitude Endurance Unmanned Aerial Vehicle Program*, Santa Monica, Calif.: RAND Corporation, MR-1054-DARPA, 1999. As of November 16, 2009:
http://www.rand.org/pubs/monograph_reports/MR1054/

Drezner, Jeffrey A., and Meilinda Huang, *On Prototyping: Lessons from RAND Research*, Santa Monica, Calif.: RAND Corporation, OP-267-OSD, 2009. As of December 15, 2009:
http://www.rand.org/pubs/occasional_papers/OP267

Executive Office of the President Office of Management and Budget, *The President's Management Agenda*, FY 2002.

Fairhall, James, "The Case for the $435 Hammer: Investigation of Pentagon's Procurement," *Washington Monthly*, January 1987.

Gansler Commission Report—*See* Commission on Army Acquisition and Program Management in Expeditionary Operations, 2007.

GAO—*See* U.S. Government Accountability Office.

Garcia, A., H. Keyner, T. Robillard, and M. Van Mullekom, "The Defense Acquisition Workforce Improvement Act: Five Years Later," *Acquisition Review Quarterly*, Vol. 4, No. 3, 1997, p. 295.

Gates, Susan M., Christine Eibner, and Edward G. Keating, *Civilian Workforce Planning in the Department of Defense: Different Levels, Different Roles*, Santa Monica, Calif.: RAND Corporation, MG-449-OSD, 2006. As of November 16, 2009:
http://www.rand.org/pubs/monographs/MG449/

Gates, Susan M., Edward G. Keating, Adria Jewell, Lindsay Daugherty, Bryan Tysinger, and Ralph Masi, *The Defense Acquisition Workforce: An Analysis of Personnel Trends Relevant to Policy, 1993–2006*, Santa Monica, Calif.: RAND Corporation, TR-572-OSD, 2008. As of November 11, 2008:
http://www.rand.org/pubs/technical_reports/TR572/

General Dynamics Electric Boat, *The Virginia Class Submarine Program: A Case Study*, Groton, Conn., February 2002.

Glennan, Thomas K., Jr., Susan J. Bodilly, Frank Camm, Kenneth R. Mayer, and Timothy J. Webb, *Barriers to Managing Risk in Large Scale Weapon System Development Programs*, Santa Monica, Calif.: RAND Corporation, MR-248-AF, 1993. As of November 16, 2009:
http://www.rand.org/pubs/monograph_reports/MR248/

GlobalSecurity.org, *LHA-1 Tarawa class Program.* As of November, 2008:
    http://www.globalsecurity.org/military/systems/ship/lha-1-program.htm

Hanks, Christopher H., Elliot I. Axelband, Shuna Lindsay, Mohammed
    Rehan Malik, and Brett D. Steele, *Reexamining Military Acquisition
    Reform: Are We There Yet?* Santa Monica, Calif.: RAND Corporation,
    MG-291-A, 2005. As of November 16, 2009:
    http://www.rand.org/pubs/monographs/MG291/

Hedgpeth, Dana, "Pentagon to Expand Its Acquisition Force," *Washington
    Post.* As of April 7, 2009:
    http://voices.washingtonpost.com/government-inc/2009/04/pentagon_
    to_expand_its_acquisi.html?wprss=government-inc

Johnson, Robert V., and John L. Birkler, *Three Programs and Ten Crite-
    ria: Evaluating and Improving Acquisition Program Management and
    Oversight Processes Within the Department of Defense,* Santa Monica,
    Calif.: RAND Corporation, MR-758-OSD, 1996. As of November
    16, 2009:
    http://www.rand.org/pubs/monograph_reports/MR758/

Klein, B. H., T. K. Glennan, Jr., and G. H. Shubert, *The Role of Prototypes
    in Development,* Santa Monica, Calif.: RAND Corporation, RM-
    3467/1-PR, February 1963. As of November 16, 2009:
    http://www.rand.org/pubs/research_memoranda/RM3467.1/

Klein, B. H., W. H Meckling, and E. G. Mesthene, *Military Research and
    Development Policies,* Santa Monica, Calif.: RAND Corporation,
    R-333, December 4, 1958. As of November 16, 2009:
    http://www.rand.org/pubs/reports/R333/

Lardner, Richard, "Army Overhauls Wartime Purchasing," *USA Today.* As of
    February 28, 2008: http://www.usatoday.com/news/washington/2008-
    02-28-2650652096_x.htm

Leonard, Robert S., Jeffrey A. Drezner, and Geoffrey Sommer, *The Arsenal
    Ship: Acquisition Process Experience,* Santa Monica, Calif.: RAND Cor-
    poration, MR-1030-DARPA, 1999. As of November 16, 2009:
    http://www.rand.org/pubs/monograph_reports/MR1030/

Lorell, Mark A., *Bomber R&D Since 1945: The Role of Experience,* Santa Mon-
    ica, Calif.: RAND Corporation, MR-670-AF, 1995. As of November 16,
    2009:
    http://www.rand.org/pubs/monograph_reports/MR670/

Lorell, Mark A., Julia F. Lowell, and Obaid Younossi, *Evolutionary Acquisition: Implementation Challenges for Defense Space Programs*, Santa Monica, Calif.: RAND Corporation, MG-431-AF, 2006. As of November 16, 2009: http://www.rand.org/pubs/monographs/MG431/

Lumb, Mark, "Where Defense Acquisition Is Today: A Close Examination of Structures and Capabilities," *Defense AT&L*, January–February, 2008.

Lynn, William, written testimony of William Lynn, Deputy Secretary of Defense, on his confirmation hearing before the Senate Armed Services Committee, Washington, D.C., January 15, 2009.

Mankins, John C., *Technology Readiness Levels: A White Paper*, NASA, Office of Space Access and Technology, Advanced Concepts Office, April 6, 1995. As of May 21, 2009: http://www.hq.nasa.gov/office/codeq/trl/trl.pdf

Margolis, M. A., R. G. Bonesteele, and J. L. Wilson, "A Method for Analyzing Competitive, Dual Source Production Programs," presented at the 19th Annual DoD Cost Analysis Symposium, September 1985.

Northrup Grumman Corporation, *Ingalls Operations: A Chronological Perspective*. As of November 8, 2008: http://www.ss.northropgrumman.com/company/chronological.html

Nunn, Sam, "Statement of Senator Sam Nunn," *Conference Report*, Vol. 132, No. 121, 1985, p. 10.

Perry, Robert L., D. DiSalvo, George R. Hall, Alvin J. Harman, G. S. Levenson, Giles K. Smith, and James P. Stucker, *System Acquisition Experience*, Santa Monica, Calif.: RAND Corporation, RM-6072-PR, November 1969. As of November 16, 2009: http://www.rand.org/pubs/research_memoranda/RM6072/

———, *A Prototype Strategy for Aircraft Development*, RAND, RM-5597-1-PR, July 1972.

Rich, Michael, and Edmund Dews with C. L. Batten, Jr., *Improving the Military Acquisition Process*, Santa Monica, Calif.: RAND Corporation, R-3373-AF/RC, February 1986. As of November 16, 2009: http://www.rand.org/pubs/reports/R3373/

Rostker, Bernard, *A Call to Revitalize the Engines of Government*, Santa Monica, CA: RAND Corporation, OP-240-OSD, 2008. As of November 16, 2009: http://www.rand.org/pubs/occasional_papers/OP240/

Sauser, B. J., J. E. Ramirez-Marquez, R. B. Magnaye, and W. Tan, *System Maturity Indices for Decision Support in the Defense Acquisition Process*, proceedings of the 5th Annual Acquisition Research Symposium, Naval Postgraduate School, Monterey, Calif., May 2008.

Schank, John F., Giles K. Smith, John L. Birkler, Brien Alkire, Michael Boito, Gordon T. Lee, Raj Raman, and John Ablard, *Acquisition and Competition Strategy Options for the DD(X): The Navy's 21st Century Destroyer*, Santa Monica, Calif.: RAND Corporation, MG-259/1-NAVY, 2006. As of November 16, 2009:
http://www.rand.org/pubs/monographs/MG259.1/

Schay, B. et al. *Summative Evaluation for DoD Labs*, Washington, D.C.: Office of Personnel Management, 2002.

Section 814 Report—*See* Defense Acquisition University, 2007.

Section 1423 Report—*See* Acquisition Advisory Panel, 2007.

Smith, Giles K., A. A. Barbour, T. L. McNaugher, M. D. Rich, and W. L. Stanley, *The Use of Prototypes in Weapon System Development*, Santa Monica, Calif.: RAND Corporation, R-2345-AF, March 1981. As of November 16, 2009:
http://www.rand.org/pubs/reports/R2345/

Smith, Giles K., and E. T. Friedmann, *An Analysis of Weapon System Acquisition Intervals, Past and Present*, Santa Monica, Calif.: RAND Corporation, R-2605-DR&E/AF, November 1980. As of November 16, 2009:
http://www.rand.org/pubs/reports/R2605/

Smith, Giles K., Hyman L. Shulman, and Robert S. Leonard, *Application of F-117 Acquisition Strategy to Other Programs in the New Acquisition Environment*, Santa Monica, Calif.: RAND Corporation, MR-749-AF, 1996. As of November 16, 2009:
http://www.rand.org/pubs/monograph_reports/MR749/

Sommer, Geoffrey, Giles K. Smith, John L. Birkler, and James R. Chiesa, *The Global Hawk Unmanned Aerial Vehicle Acquisition Process*, Santa Monica, Calif.: RAND Corporation, MR-809-DARPA, 1997. As of November 5, 2009:
http://www.rand.org/pubs/monograph_reports/MR809

Thirtle, Michael R., Robert V. Johnson, and John L. Birkler, *The Predator ACTD: A Case Study for Transition Planning to the Formal Acquisition*

Process, Santa Monica, Calif.: RAND Corporation, MR-899-OSD, 1997. As of November 16, 2009:
http://www.rand.org/pubs/monograph_reports/MR899/

U.S. Department of Defense, DoD Instruction 5000.55, "Reporting Management Information on DoD Military and Civilian Acquisition Personnel and Position," November 1, 1991.

———, *Technology Readiness Assessment (TRA) Deskbook*, prepared by the Deputy Under Secretary of Defense for Science and Technology (DUSD (S&T)), Washington, D.C., May 2005. As of May 21, 2009:
http://www.dod.mil/ddre/doc/May2005_TRA_2005_DoD.pdf

———, *Manager's Guide to Technology Transition in an Evolutionary Acquisition Environment*, Version 2.0, Fort Belvoir, Va.: Defense Acquisition University Press, June 2005.

———, DoD Instruction 5000.66, "Operation of the Defense Acquisition, Technology, and Logistics Workforce Education, Training, and Career Development Program," December 21, 2005.

———, *Quadrennial Defense Review Report*, February 6, 2006. As of April 7, 2009:
http://www.defenselink.mil/qdr/report/Report20060203.pdf

———, DoD Instruction 5000.2, "Operation of the Defense Acquisition System," Washington, D.C., December 2, 2008.

———, DoD Instruction 5000.2, "Operation of the Defense Acquisition System," December 8, 2008.

———, "DoD and OPM to Review National Security Personnel System." As of March 16, 2009:
http://www.defenselink.mil/releases/release.aspx?releaseid=12556.

U.S. Department of Defense, Acquisition, Technology, and Logistics, *AT&L Human Capital Strategic Plan*, Volume 1, 2006.

———, *AT&L Human Capital Strategic Plan*, Version 3.0, 2007.

U.S. Department of Defense, Office of the Inspector General, *Human Capital: Report on the DoD Acquisition Workforce Count*, D-2006-073, 2006.

U.S. Department of Defense, Under Secretary of Defense, Acquisition, Technology, and Logistics, DoD Instruction 5000.02, "Operation of the Defense Acquisition System," Washington, D.C., December 2, 2008.

U.S. General Accounting Office, *Acquisition Workforce: Department of Defense's Plans to Address Workforce Size and Structure Challenges*, GAO-02-630, April 2002.

U.S. Government Accountability Office, *Defense Acquisitions: Challenges Facing the DD(X) Destroyer Program*, GAO-04-973, September 2004.

———, *Defense Acquisitions: Progress and Challenges Facing the DD(X) Surface Combatant Program*, Paul Francis, "Testimony Before the Subcommittee on Projection Forces, Committee on Armed Services, House of Representatives," GAO-05-924T, July 19, 2005.

———, *Defense Acquisitions: Challenges Remain in developing Capabilities for Naval Surface Fire Support*, GAO-07-115, November 2006.

———, *Success of Advanced SEAL Delivery System (ASDS) Hinges on Establishing a Sound Contracting Strategy and Performance Criteria*, Washington, D.C., GAO-07-745, May 2007.

———, "Defense Acquisitions: Realistic Business Cases Needed to Execute Navy Shipbuilding Programs, Statement of Paul L. Francis, Director Acquisition and Sourcing Management Team, Testimony Before the Subcommittee on Seapower and Expeditionary Forces, Committee on Armed Services, House of Representatives," Washington, D.C., GAO-07-943T, July 24, 2007.

———, *Defense Acquisitions: Assessment of Selected Weapon Programs*, GAO-08-467-SP, March 2008.

———, *Defense Acquisitions: Assessment of Selected Weapon Programs*, GAO-09-326-SP, March 2009a.

———, *Department of Defense: Additional Actions and Data are Needed to Effectively Manage and Oversee DoD's Acquisition Workforce*, GAO-09-342, March 2009b.

U.S. Navy, "Statement of the Honorable Dr. Delores M. Etter, Assistant Secretary of the Navy (Research, Development, and Acquisition); VADM Paul E. Sullivan, U.S. Navy, Commander, Naval Sea Systems Command; RADM Charles S. Hamilton, II, U.S. Navy, Program Executive Officer, Ships; and RADM Barry J. McCullough, U.S. Navy Director of Surface Warfare, Before the Subcommittee on Seapower and Expeditionary Forces of the House Armed Services Committee on Acquisition Oversight of the U.S. Navy's Littoral Combat Ship Program," Washington, D.C., February 8, 2007.

U.S. Office of Personnel Management, *FERS—Federal Employees Retirement System Transfer Handbook: A Guide to Making Your Decision*, RI 90-3, 1997. As of July 5, 2007:
http://www.opm.gov/retire/fers_election/fersh/hb.pdf
————, *Handbook of Occupational Groups and Families*, January 2008.

Vernez, Georges, and H. Garrison Massey, *The Acquisition Cost Estimating Workforce: Census and Characteristics*, Santa Monica, Calif.: RAND Corporation, TR-708-AF, 2009.

Vernez, Georges, Albert A. Robbert, Hugh G. Massey, and Kevin Driscoll, *Workforce Planning and Development Processes: A Practical Guide*, Santa Monica, Calif.: RAND Corporation, TR-408-AF, 2007.
http://www.rand.org/pubs/technical_reports/TR408/

Woolner, Derek, *Getting in Early: Lessons of the Collins Submarine Program for Improved Oversight of Defence Procurement*, Parliament of Australia, Parliamentary Library, 2001. As of September 15, 2009:
http://www.aph.gov.au/library/pubs/rp/2001-02/02RP03.htm

Young, John J., Under Secretary of Defense (Acquisition, Technology and Logistics), "Prototyping and Competition," Policy Memorandum, September 19, 2007.

Younossi, Obaid, Mark V. Arena, Robert S. Leonard, Charles Robert Roll, Arvind K. Jain, Jerry M. Sollinger, *Is Weapon System Cost Growth Increasing? A Quantitative Assessment of Completed and Ongoing Programs*, Santa Monica, Calif.: RAND Corporation, MG-588-AF, 2007. As of November 16, 2009:
http://www.rand.org/pubs/monographs/MG588/

Younossi, Obaid, David E. Stem, Mark A. Lorell, and Frances M. Lussier, *F/A-22 and F/1-18E/F Development Programs*, Santa Monica, Calif.: RAND Corporation, MG-276-AF, 2005. As of November 16, 2009:
http://www.rand.org/pubs/monographs/MG276/